The Politics of Riverine Rights achieves a rare feat, exemplarily analyzing specific cases while generating informed critique beyond them. It should be essential reading for anyone interested in a balanced approach to the rights of nature, and generally in how to conduct excellent interdisciplinary research in times of environmental undoing.

Professor Mihnea Tănăsescu, National Fund for Scientific Research (FNRS) and University of Mons, Belgium, author of *Understanding the rights of nature: A critical introduction.*

Exploring the nature and effects of granting legal personhood to rivers in three landmark cases, this volume offers rich, empirically grounded reflections on the possibilities for more relational understandings of human and more-than-human rights.

Rachel Sieder, Centre for Research and Graduate Studies in Social Anthropology (CIESAS), Mexico City, co-editor of *The handbook of law and society in Latin America.*

Through several illuminating case studies, this book offers a deep and sophisticated analysis of the riverine rights movement. It is essential reading for understanding this growing field and how it can continue to evolve toward stronger enforcement and broader impact. Perfect for policymakers, academics, activists, and curious readers alike.

Grant Wilson, Executive director, Earth Law Center, USA, co-editor of *Earth Law: Emerging Ecocentric Law—A Guide for Practitioners.*

The Politics of Riverine Rights

This edited collection is based on comparative ethnographic research and explores the concept of riverine rights in Aotaroa New Zealand, Colombia and India where rivers have been declared legal persons: the Whanganui River, Río Atrato, the Ganga and the Yamuna.

The cases drew significant attention from academia and the wider public, as examples of the rights of nature in practice. This book contributes with in-depth empirical research carried out by a highly interdisciplinary team of co-authors – from anthropology, law, history, political science, management studies and water science – many of whom are from the case study countries and the river basins. The first chapters carefully examine each case, discussing its history, the institutional context, the actors and interests, the issues to be addressed by the legal decisions and their outcomes. Subsequent chapters provide detailed comparative analysis on the rights of nature legal pluralism, the struggles of Indigenous peoples, ethnic minorities and marginalized groups, and on different ontological understandings of the relations between people, rivers and nature. The book concludes with reflections on how these cases of riverine rights relate to global trends, and discusses the tensions and synergies between human rights and nature's rights, with a particular emphasis on how Indigenous rights intersect with rivers' rights. Finally, it argues for a broad notion of *riverine* rights that includes those who live by the river, and *beyond* riverine rights, to the social, cultural and political dimensions of the legal innovations, and to the larger environmental crises they seek to address.

This interdisciplinary book will be essential reading for academics and researchers across social sciences and law with an interest in how social struggles, environmental justice and innovative legislation intersect for the wellbeing of rivers and their people. The book is also relevant for policy makers, activists and others with an interest in rights of nature.

Malene K. Brandshaug is a postdoctoral fellow at the Department of International Studies, Oslo Metropolitan University, Norway, and holds a PhD in social anthropology from the School of Global Studies at University of Gothenburg (2020), Sweden. She has a decade-long ethnographic specialization in water and Peru, has recently worked on human-water relations and environmental activism in Norway and on international currents regarding rights of rivers. Her current postdoctoral project is titled Rights of Waters: An Ethnographic Study of Human-Environment Relations and Local Mobilisation in Norway and is part of the research project Riverine Rights: Exploring the Currents and Consequences of Legal Innovations on the Rights of Rivers.

Axel Borchgrevink is Professor of development studies at Oslo Metropolitan University, Norway. He holds a PhD in social anthropology from the University of Oslo. He has done research in Latin America, the Horn of Africa, the Philippines and New Zealand, and worked with issues of rural development and resource politics, civil society and Indigenous rights, and development aid. He is the author of *Clean and Green: Knowledge and Morality in a Philippine Farming Community* and has also co-edited the book *Contested Powers: The Politics of Energy and Development in Latin America*. He is the project leader of Riverine Rights: Exploring the Currents and Consequences of Legal Innovations on the Rights of Rivers.

Routledge Studies in Environmental Justice

This series is theoretically and geographically broad in scope, seeking to explore the emerging debates, controversies and practical solutions within Environmental Justice, from around the globe. It offers cutting-edge perspectives at both a local and global scale, engaging with topics such as climate justice, water governance, air pollution, waste management, environmental crime, and the various intersections of the field with related disciplines.

The *Routledge Studies in Environmental Justice* series welcomes submissions that combine strong academic theory with practical applications, and as such is relevant to a global readership of students, researchers, policy-makers, practitioners and activists.

Environmental Justice in Early Victorian Literature
Adrian Tait

Guerrilla Ecologies
Green Capital, Nature, and the Politics of Catastrophe
John Maerhofer

Rectifying Climate Injustice
Reparations for Loss and Damage
Laura García-Portela

Environmental Justice in Nepal
Origins, Struggles, and Prospects
Edited by Jonathan K London, Jagannath Adhikari, and Thomas Robertson

The Conservation of Violence
Statecraft, Forests, and Coloniality
Tafadzwa Mushonga

The Politics of Riverine Rights
Environmental Struggles in Aotearoa New Zealand, Colombia and India
Edited by Malene K. Brandshaug and Axel Borchgrevink

For more information about this series, please visit: www.routledge.com/Routledge-Studies-in-Environmental-Justice/book-series/EJS

The Politics of Riverine Rights

Environmental Struggles in Aotearoa New Zealand, Colombia and India

Edited by
Malene K. Brandshaug and Axel Borchgrevink

Routledge
Taylor & Francis Group
LONDON AND NEW YORK

First published 2026
by Routledge
4 Park Square, Milton Park, Abingdon, Oxon OX14 4RN

and by Routledge
605 Third Avenue, New York, NY 10158

Routledge is an imprint of the Taylor & Francis Group, an informa business

British Library Cataloguing-in-Publication Data
A catalogue record for this book is available from the British Library

ISBN: 978-1-032-94678-8 (hbk)
ISBN: 978-1-032-94679-5 (pbk)
ISBN: 978-1-003-58120-8 (ebk)

DOI: 10.4324/9781003581208

Typeset in Times New Roman
by Taylor & Francis Books

Contents

Illustrations

Figures

Tables

Box

Contributors

Axel Borchgrevink is Professor Emeritus of development studies at Oslo Metropolitan University, Norway. He holds a PhD in social anthropology and has done research in Latin America, the Horn of Africa, the Philippines and New Zealand on issues of rural development and resource politics, civil society and indigenous rights, and development aid.

Malene K. Brandshaug is a postdoctoral researcher in development studies at Oslo Metropolitan University, Norway, and holds a PhD in social anthropology from the School of Global Studies at Gothenburg University, Sweden. She has done ethnographic research on environmental struggles and human-water relations in Peru and Norway.

Miriama Cribb is a born and bred descendant of the tribes of the Whanganui River. She has affiliations to other tribes but is most active at her home in Whanganui. Miriama is completing a PhD on implementing Indigenous frameworks in non-Indigenous organisations, using Te Awa Tupua as her case study.

Camilo Arango Duque is a postdoctoral researcher in legal theory at EAFIT University, Colombia. He holds a Ph.D. in humanities and an LLM in environmental law. His research interests include environmental law and environmental legal history. He is academic reporter for the Yearbook of International Environmental Law of Oxford University Press.

Daniel Alfonso León is a historian with a master's degree in history from the Universidad Industrial de Santander (UIS), Colombia. He specializes in energy and society in contemporary capitalism at the Federal University of Rio de Janeiro (UFRJ) and is a doctoral candidate in the Latin American Integration Program at the University of São Paulo (USP).

John Andrew McNeish is a social anthropologist and Professor of international environment and development studies at the Norwegian University of Life Sciences. His research and publications focus on a series of interlinked fields stretching from Indigenous people's rights and politics, development and security policy, and natural resource and energy politics.

Julián Gaviria Mira is assistant professor of legal theory at EAFIT University, Colombia. He holds a Ph.D. in law and a Master in public law from Universidad Carlos III de Madrid, and a Master in constitutional law from CEPC, Madrid. His research focuses on constitutional theory, democracy and political inclusion.

Nataly Montoya-Restrepo is assistant professor in the Department of Public Law at EAFIT University, Colombia. Her research focuses on urban planning, critical and feminist geography, constitutional law and ecological transitional justice. She holds a Ph.D. in geography from Pontificia Universidad Católica, Chile and a Master in urban-regional studies from Universidad Nacional de Colombia.

Arlinton Cuesta Mosquera is a lawyer from the Universidad Tecnológica del Chocó with expertise in human rights land restitution and the defence of vulnerable ethnic communities in Colombia. He has worked with legal proceedings, public policy and collective reparation, and holds a master's degree in education and human rights from UNAULA, Colombia.

Catalina Vallejo Piedrahíta is assistant professor in the Department of International Law at EAFIT University, Colombia. Her research interests include climate litigation, environmental law and ecological approaches to transitional justice. She holds a Ph.D. in law, and a M.A. in peace studies.

Jenny Moreno Socha is a political scientist with a master's degree in political studies from the National University of Colombia. She specializes in public policies for equality in the Latin American Council of Social Sciences and is a doctoral candidate in the Latin American Integration Program at the University of São Paulo.

Kavita Upadhyay is an independent journalist and researcher whose work lies at the intersection of policy, politics and the environment in the Indian Himalayan Region. She has worked as a research associate on the project Riverine Rights: Exploring the Currents and Consequences of Legal Innovations on the Rights of Rivers.

Acknowledgments

Thanks to the people of Whanganui, Atrato and Uttarakhand! There are too many of you to acknowledge individually, but you know who you are. This book would not have come about if it wasn't for the generosity and openness with which you received us, and for your willingness to share your reflections with us.

Thanks to the rivers Whanganui, Atrato, Ganga and Yamuna. It was wonderful to be able to see and experience you. We appreciate your uniqueness, yet your waters all flow into the sea and remind us of our shared environmental and social concerns and the importance of fine tuning our laws and politics to safeguard the wellbeing of rivers and riverine communities.

Thanks to Elizabeth Macpherson, Bibhu Prasad Nayak and Rahul Ranjan! You were great colleagues throughout the Riverine Rights research project. Unfortunately, none of you were able to take part in the writing of this book, which is a loss given your rich insights into the rivers, their people and the legal personhoods. But your work and your contributions to the project have been important for the writing of the book and are reflected in every page.

Thanks to all of you who have contributed with discussions and perspectives at various stages of the research process, including: Harry Blair, Rutgerd Boelens, Omaira Bolaños, Camilla Brattland, Anouk Brisebois, Cristy Clark, Sissel Egden, Eva Maria Fjellheim, Tom Griffiths, Jo Helle-Valle, Benedikte Lindskog, Martin Lee Müller, Kenneth Bo Nielsen, Velina Ninkova, Kristen Nordhaug, Erin O'Donnel, Simon Pahle, Brita Phuti, Nancy Postero, Anindita Punjari, Hege Roaldset, Cesar Rodríguez Garavito, Rachel Sieder, Hanne Svarstad, Mihnea Tanasescu, Jessica Ullevålseter, Christina Voigt, Anne Waldrop, Gro Ween and Grant Wilson.

Thanks to Oslo Metropolitan University for hosting the Riverine Rights project, and to the Norwegian University of Life Sciences, the University of Canterbury (Christchurch, Aotearoa New Zealand), Tata Institute of Social Sciences Hyderabad (India) and Universidad Autónoma Latinoamericana (Medellin, Colombia) for being partners in the undertaking. A huge thanks to the Research Council of Norway for funding the whole research project. And thanks to OsloMet, to the research group Development, Power and Inequality and to the Royal Norwegian Society of Sciences and Letters for grants to make this book available as Open Access.

1 Exploring riverine rights

Axel Borchgrevink and Malene K. Brandshaug

Introduction

Rivers all around the world are subject to environmental degradation and threats to their health and existence. The lives of humans, animals and plants living along the rivers are also affected. Many riverine communities care for and fight for the wellbeing of their river. There are also environmental movements concerned by the state of rivers and other water bodies which mobilize to secure their protection. Although environmental laws and regulations to protect rivers exist in most places, this has not been sufficient to safeguard them. In response to this, a growing number of rivers are given legal rights. This approach is part of a larger trend of granting rights to nature. Over a few months – November 2016 to March 2017 – rivers in Colombia, Aotearoa New Zealand and India were all recognized as legal persons or legal subjects, with corresponding rights. This book is about these rivers and their people, and in particular about understanding the circumstances and implications of granting them rights.

On November 10, 2016, the Constitutional Court in Colombia issued *Sentencia* T-622, which declared the Atrato River (Río Atrato) to be a subject of rights, and introduced the concept of "biocultural rights", to be accorded jointly to the river and the people living along its banks and depending on the river. A group of guardians to act on behalf of the river was named, including representatives from river communities. The ruling also contained orders to address and resolve the serious contamination of the river due to illegal gold extraction. The ruling was the conclusion to a case initiated by a constitutional writ of petition – a case of public interest litigation[1] – brought by the Afro-Colombian communities of the river for state intervention to protect their human rights, which were infringed upon because of the contamination and deforestation caused by legal and illegal gold mining in the river basin.

On March 20, 2017, Parliament of Aotearoa New Zealand passed the *Te Awa Tupua* (Whanganui River Claims Settlement) Act, which declared the Whanganui River to be a legal person, and further stated that "Te Awa Tupua is an indivisible and living whole, comprising the Whanganui River from the mountains to the sea, incorporating all its physical and metaphysical

DOI: 10.4324/9781003581208-1

elements" (Te Awa Tupua Act 2017, s.12). In this way, the state concluded the process of repairing the injustice against the Whanganui Māori *iwi* (tribe) of not respecting their sovereignty over the river from the time of colonization. The new legislation gave Māori law (*kawa*) a prominent place to ensure that the future management would deal with the river in a holistic way and prioritize the river's health. It also established institutions to secure Māori influence.

Also on March 20, 2017, the High Court of Uttarakhand State in India declared the Ganga (Ganges) and Yamuna rivers to be living entities "with all rights, duties and liabilities of a living person" (Mohd Salim v. State of Uttarakhand 2017, n 19). Ten days later, on March 30, the same bench of Uttarakhand High Court also declared the glaciers that feed these two rivers to be legal persons. These rulings were also resulting from public interest litigation, after two individuals had petitioned the court to oblige the state to address, respectively, illegal constructions and encroachments on the Ganga, and the pollution of the Ganga from sewage and industrial effluents. A group of state officials were named as guardians to represent the rivers and ensure their rights. However, the state of Uttarakhand challenged the rulings to the Supreme Court in India, which stayed them in July 2017. They therefore remain suspended until the Supreme Court decides on the matter, which is not expected to happen in the near future.

These cases of "river persons"– the Atrato River, the Whanganui River, and the Ganga (Ganges) and Yamuna Rivers – received widespread international attention. Much media reporting focused on the novelty of framing a river as a person, a person that was able to take its complaints to court. The idea was startling and counterintuitive for many, and there were widespread comments making fun of it or dismissing it as another woke phenomenon. But these legal decisions were also hailed as legal innovations which open up a new way of protecting the environment by granting rights to nature. These cases of "riverine rights"– especially the first two – have become emblematic examples of a rapidly growing movement for the promotion of the rights of nature. Giving legal rights to nature, it is claimed, will lead to better forms of environmental protection and governance, so urgently needed in the era of the Anthropocene, where human activity is disturbing the climate and eroding biodiversity.

When the researchers behind this book learnt of these cases, we were intrigued by a number of things. The idea of seeing a natural entity such as a river on par with a person, at least in legal terms, appeared to be a radical break with formal systems of state law. It gave rise to a host of questions.

One set of questions centred on why: Why did this legal innovation happen in three so very different countries at more or less the same time? What kinds of motives, drivers and processes were behind these three cases? Why now? Why were rivers the natural entity that was given personhood?

Another set of questions focused on outcomes: Was legal personhood for rivers compatible with other legal principles, such as private ownership to

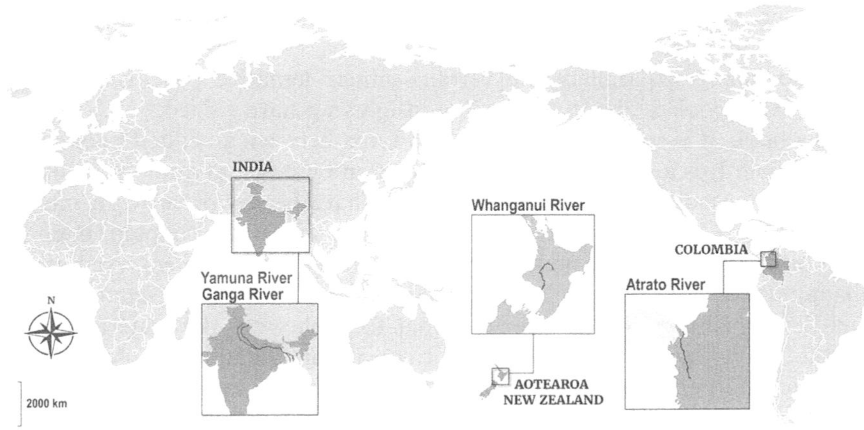

Figure 1.1 Aotearoa New Zealand, Colombia and India – on three different continents. Map data ©2025 Google, INEGI. Elaborated by Daniel Alfonso León.

land? Could rights of nature come into conflict with human rights? Would these innovations have any effects on environmental governance and protection? Or would they remain as symbolic expressions of good intentions without any practical impact, as paper tigers?

And a third set of questions we had related to what we can learn from these cases. Are there lessons from the three cases that can be useful for other contexts? Are they examples of overcoming the destructive patterns in society's way of interacting with the natural environment? Can they throw light on broader theoretical debates over how social science should study humans and their relationship with their surroundings?

This book is the product of a research project designed to investigate these issues. Between 2020 and 2024, an international group of researchers studied the cases and their implications. Interdisciplinary and contextual empirical research combined ethnographic fieldwork, interviews and written sources to understand each case in as much detail and from as many angles as possible. The project group included researchers from each of the three countries, and from different disciplines, including anthropology, law, political science, history, environmental economics and organizational studies. In the research, we worked with and supported local grassroots and Indigenous organizations. In order to facilitate comparison and learning between the cases, the core research group jointly visited all rivers.

Working in this collaborative way and thereby getting to know about these cases and the societies they belong to has been a great learning experience for all of us. One thing we quickly found out was that what at first sight appeared to be three instances of the same thing – legal personhood for a river – upon closer examination turned out to be wildly different. And different not only along one axis, but different in terms of scale, legal form, popular involvement, drivers of the cases, effects and more. Yet, as we also found out, there are also linkages, parallels and certain similar tentative conclusions to be drawn. This book is about sharing the insights we have gained.

As a book, it is something of a hybrid: a mix between an edited book and a monograph. It is a collection of articles written by different authors, so in this sense it is an edited book. But all chapters have been written to fit into an outline with a logic like that of a monograph, where the chapters build on each other and gradually make a more comprehensive argument. And all the chapters have authors who have worked together on this project over years and developed common perspectives that form the way the material is presented. While this should not be taken to mean that all authors hold the same opinions and perspectives, it does mean that it has been possible to construct a more coherent book than what is often the case with edited volumes.

In the coming chapters, we examine closely the context and background for why these rivers were given legal personhood, and we discuss the implications of the decisions to grant legal personhood. As will appear, rights of nature is by no means the only lens or perspective through which to understand these cases. Indeed, in part we argue that there are other frames that may be equally or more relevant for understanding important aspects. Yet, the international literature on the cases exclusively discusses them in relation to rights of nature. A general presentation of the ideas of rights of nature, and the different strands that make up this field, is therefore necessary as a background against which the argument of this book should be read.

In very broad strokes, the literature on rights of nature can be divided into two camps. There are those writers for whom rights of nature are inherently a good thing, an idea that should be promoted. The title of Boyd's 2017 book, *The rights of nature: A legal revolution that could save the world*, neatly summarizes this stance (other examples include Kauffman and Martin 2021, Vega Cárdenas and Turp 2023 and García Ruales et al. 2024). And then there are those that come to the field without a preformed conclusion but seek to assess the merits and difficulties of rights of nature in a dispassionate and critical way (e.g., O'Donnell 2018, Macpherson 2019, Tanasescu 2022). This book belongs in the latter camp. We have been more interested in understanding what is happening in our cases by analysing them in their contexts, and we have come to the rivers with quite open minds as to what we would find. The book reflects this open and exploratory approach.

In what follows of this introduction, we give a general presentation of the ideas of rights of nature, and of what we have termed riverine rights, and of some of the questions and debates surrounding them. Thereafter we present our research project. In particular, we outline the methodological choices we have made. This is important background for assessing the following chapters, and we also think the design, approach and organization of the research can be useful as an example of a comparative study of this type. We end this opening chapter with an overview of the remainder of the book, and the logic behind it.

Rights of nature, personhood rights and riverine rights

Rights of nature

While the idea of rights of nature is still unfamiliar to many, there are good reasons for seeing this as a growing phenomenon. Since the first examples of rights of nature jurisprudence less than 20 years ago, there are now around 500 initiatives for legal provisions centring on nature that have been registered by the Eco Jurisprudence Monitor.[2] Graphs of the rise of such cases show steady growth over the years. These initiatives are diverse. They are introduced in different forms, from municipal ordinances to constitutional provisions, and stem from court decisions as well as elected assemblies at different levels. Their reach and impact vary greatly. But taken together, these examples can be seen as an expression of a global movement. This movement has been spearheaded by environmental lawyers and certain environmental non-governmental organizations. The Global Alliance for the Rights of Nature (GARN) is an umbrella organization that unites many of them. The movement is not limited to civil society, though. In addition to the growing number of rights of nature within state law, the United Nations is also developing similar principles in its process for *Harmony with Nature.*

Several different terms are used more or less interchangeably with rights of nature – such as earth law, ecocentric jurisprudence and wild law. A common claim is that the new approach breaks with traditional *anthropocentric* law – law that aims to protect and promote the interests of humans. Instead, rights of nature is called an *ecocentric* approach, one that has the protection of the environment as an objective in its own right. Where human rights laws typically protect the rights of humans and environmental laws typically protect the environment on the basis of human interest, the novelty of rights of nature laws lies in their positioning of nature as the very subject of rights. In this way, the proponents of rights of nature see the new approach as breaking with the mainstream view of nature as a resource for the use of humans, the value of which stems from the potential benefits we humans may have from it. Instead, the approach is based on recognizing the inherent value of nature in itself. In this way, the ecocentric approach of rights of nature can imply a fundamental break with the way that our economic system is exploiting and degrading the environment, it is argued. This opposition between anthropocentrism and ecocentrism is fundamental within the discourse of rights of nature. In several places throughout this book, we will discuss the relevance of this opposition for the field.

The standard reference when talking about the history of rights of nature is Christopher Stone and his 1972 article “Should trees have standing?”. Stone’s point of departure is a court case that the conservationist organization The Sierra Club filed against the development of a tourist project in the Sequoia National Forest in California, on the grounds that the project would harm the ecological balance. However, the case was dismissed by the court on the grounds that the Sierra Club did not have legal standing to bring the suit. The Supreme Court upheld this decision. The case spurred Stone to write his seminal article, where he argues that the history of law shows an ever-expanding circle of subjects that are accorded rights – he mentions prisoners, aliens, women (especially of the married variety), the insane, Blacks, foetuses and Indians as examples of categories of persons that increasingly (although imperfectly) have been accorded rights (Stone 1972, 451). Moreover, he points out, also non-animate rights-holders have been acknowledged by the law: trusts, corporations, municipalities, nation states and even ships are recognized as legal persons with rights (Stone 1972, 452). He goes on to argue that each of these expansions of right-holders appeared as jarring and unthinkable up until the point when rights were accorded. Thus, he argues, there is nothing different in expanding rights to also include rivers, forests, lakes, mountains or other natural entities – or to nature itself:

> Of course, to urge a court that an endangered river is 'a person' … will call for lawyers as bold and imaginative as those that [in 1886] convinced the Supreme Court that a railroad corporation was a person under the fourteenth amendment, a constitutional provision theretofore generally thought of as designed to secure the rights of freedmen.
>
> (Stone 1972, 465)[3]

According to Stone, in 1972 when he was writing his article, the time was ripe for bold and imaginative lawyers to take the next step and give rights to nature. He argued that this would not only allow courts to take on cases of damage to nature (give standing) but would also allow damages to be assessed in terms of damage to the natural entity (and not losses to affected persons), and those damages to be used for restoration. For such rights to have any reality, there would need to be living persons to speak and act on behalf of the natural entity "as lawyers speak on behalf of corporations" (Stone 1972, 464). Stone suggests a system in which "a friend of a natural object [who] perceives it to be endangered, … can apply to the court for a guardianship to be established" (Stone 1972, 464).

Stone's article was prescient in bringing out key elements that have remained central in the establishment of rights of nature over the past two decades. It is no wonder that it is often referred to in what Mihnea Tanasescu (2022) has termed the origin story of rights of nature. But, as Tanasescu demonstrates, there are also other strands to the history of the idea that help explain other elements of the current movement for rights of nature. While Stone was basically a pragmatic, arguing for rights of nature as a way of securing desired outcomes, there are other proponents of the idea for whom the crucial aspect of rights of nature is that it is a recognition of the intrinsic value of nature. For them, recognizing the rights of nature may be just as much a moral revolution – breaking with "anthropocentric ethics"– as a legal innovation. Tanasescu traces these ideas from writers such as Godofredo

Stutzin (1984), Thomas Berry (2010) and Cormac Cullinan (2011), who have all been influential in the development and spread of the idea of the rights of nature. Due to the way that religious convictions underpin the perspectives of these writers, Tanasescu refers to this strand of the rights of nature movement as ecotheology. However, as it is possible to argue that the idea of nature's intrinsic value also has non-religious roots – such as the "deep ecology" of Arne Næss (Naess 1989) – this term may be somewhat misleading.

Another prominent feature of the discourse of rights of nature is the connection made with Indigenous peoples and their lifeworlds. The idea of rights of nature is said to express Indigenous thinking. In the constitution of GARN, for instance, it is stated that one of its objectives is

> to honor the wisdom of Indigenous Peoples, whose laws are drawn from their inherent relationship with Mother Earth, and strengthen Indigenous Peoples' perspectives within the Rights of Nature, Rights of Mother Earth and Earth Laws movement.
>
> (GARN 2022, 1)

Many of the rights of nature examples that exist have strong links to Indigenous peoples, including landmark cases such as the Ecuadorian constitution (see below) and the cases from Aotearoa New Zealand and Colombia that we examine in this book. Yet this identification of rights of nature with Indigenous philosophy may run the risk of simplifying and essentializing, both by postulating a universal Indigenous thought and by the implicit opposition of the Indigenous to the Western. Tanasescu even argues that identifying Indigenous thinking with a right of nature-perspective misrepresents much Indigenous thought and underestimates its radical potentials (2022, see also Marshall 2019). The relationship between Indigenous rights and rights of nature is one of the issues that we will be focusing on in this book.

Personhood rights

A milestone in the history of rights of nature was the 2008 Constitution of Ecuador. Its article 71 states:

> Nature, or Pachamama, where life is reproduced and occurs, has the right to integral respect for its existence and for the maintenance and regeneration of its life cycles, structure, functions and evolutionary processes. All persons, communities, peoples and nations can demand public authorities enforce the rights of nature.

And article 72 goes on to say that "Nature has the right to be restored". Two years later, Bolivia passed similar legislation, in the form of the *Ley de Derechos de la Madre Tierra* (Law of the Rights of Mother Earth).

These two cases give rights to all of nature, or to nature as a whole, *Pachamama* or Mother Earth. In contrast, the cases we are examining in this book are examples of rights being given to specific natural entities – rivers – that are granted legal personhood. When Rio Atrato (in October 2016), the Whanganui River and the Yamuna and Ganga Rivers (all in March 2017) were granted rights, part of the novelty was precisely in singling out rivers for legal personhood.

Kauffman and Martin (2021) speak of two different models for rights of nature, which they call *the Nature's Rights Model* and *the Personhood Model.* They emphasize that these are ideal models, and that existing legal provisions may incorporate elements from both. As the names imply, the scope differs between the two models – rights for all of nature within the jurisdiction vs only for one specific ecosystem. Also, the use of legal personhood applies only in the latter case. They point to other differences between these models as well. In the personhood model, specific guardians are pointed out to represent and speak for the natural entity. Having designated guardians, they claim, may enable a more proactive realization of the rights of nature. In contrast, the nature's rights model they see as mainly coming into play when nature is already damaged or threatened by damage.

Furthermore, Kauffman and Martin argue that there are different types of rights involved in these two models. The type of rights that are given to all of nature (for instance in Ecuador and Bolivia) are sometimes termed *existential rights.* As we saw in the Ecuador case, they involve the rights to existence, maintenance, regeneration and restoration. According to Kauffman and Martin, the personhood model means extending *human rights* to the ecosystem in question (Kauffman and Martin 2021, 15). For Kauffman and Martin, this is a crucial distinction, as they argue that since the personhood model only transfers human rights to nature, it does not open for the transformation of the legal paradigm into a new way of relating with nature – which for them is the objective of rights of nature (2021, 219 and 220).

The issue of what types of rights are accorded in the different legal arrangements for rights of nature is one that we will be examining in this book. In our view, the issue is more complex than what Kauffman and Martin present. On the one hand, the personhood model does not extend the full range of human rights to a natural entity, but a selected set (just as other legal persons – such as corporations – are constructed to have certain types of rights but not all). So, the set of rights accorded to the natural entity can be tailored to the type of entity in question. Moreover, legal scholars have argued that the very notion of rights implies a rights-bearing subject – that is, some kind of person. In line with this argument, also the nature's rights model presupposes some form of personhood. The central use of the concept of "Mother Earth" (*Pachamama*) in the Ecuadorean constitution certainly implies a form of personification of nature. Tanasescu also questions the uniqueness of these existential rights granted to nature. According to him, the rights recognized for (all of) nature in the writings of Berry and Cullinan –

"the right to exist, to have a habitat, and to evolve as part of the Earth community" – are strikingly similar to the human rights discourse, and to "Thomas Jefferson's fundamental rights … to life, liberty, and the pursuit of happiness" (Tanasescu 2022, 29).

Therefore, for Tanasescu, the difference between the two models does not reside in the types of rights accorded. Instead, he points to the different conceptions of nature underlying these models as the crucial distinction between them. The nature's rights model refers to what he calls Nature with a capital N. That is, an abstracted notion of everything that surrounds us, the totality, which according to Tanasescu is at such an abstract level that it allows for no concrete relationship between humans and this Nature (with a capital N). Similarly, it places actual environments as derivative only, as somehow subservient to the totality (Tanasescu 2022, 32–34). In contrast, he claims, the personhood model, by focusing on specific environments, implies a place-based understanding of nature that is particular and allows for particular ways of relating to it. For Tanasescu (in contrast to Kauffman and Martin), it is therefore the personhood model that has the potential to be transformative, by allowing for relational understandings of humans and their surroundings that are concrete and experiential.

Our study of the three river cases throws light on these issues in various ways. We analyse how elements from the two ideal models are combined in practice, the types of rights that are realized in the cases, the roles of guardians and the underlying conceptions of nature.

Riverine rights

The cases described in this book are among the early cases of rights of nature and in particular early cases of giving personhood rights to natural entities. There was at the time only one example of the personhood model (the former national park Te Urewera in Aotearoa New Zealand in 2014). There were a few examples of rights of nature more broadly. In 2006, Tamaqua Borough in Pennsylvania recognized the rights of nature in local law. The Ecuadorean 2008 Constitution and the Bolivian Law of Mother Earth of 2010 represent the breakthrough of rights of nature in constitutions and national law. The cases we examine are the first ones that involve giving rights to *rivers*. Hence, the granting of personhood rights to Whanganui River in Aotearoa New Zealand, Atrato River in Colombia and the rivers Ganga and Yamuna in India are pioneer cases of personhood rights to rivers – a phenomenon that, as with rights of nature more generally, we only now are starting to see the outcomes of and learn from.

A concern in rights of nature debates is whether the move towards rights for nature and natural entities, such as rivers, may come into conflict with human rights by sidelining humans for the benefit of the environment. This critique has recently been challenged in rights of nature scholarship by using specific cases to show that there can be synergies between human rights and nature's rights (Tanasescu et al. 2024) and to demonstrate the centrality of Indigenous rights in specific river rights cases (Cribb et al. 2024). In this book, the cases from Aotearoa New Zealand, Colombia and India demonstrate that human matters can be integrated with riverine rights. Conversely, another concern could be the position that these cases are not really about river rights but about human rights – or, in some cases, about Indigenous rights – concealed as river rights. In this book, we are interested in exploring how different concerns, of nature and of humans, relate in each case. In the Whanganui River case, for instance, it is exactly the centring of the river that ensures that neither human nor river interests are lost out of sight, since these are intrinsically woven together. This is highlighted in the Te Awa Tupua Act as "I am the river, the river is me", a central principle in Māori Whanganui *kawa* (law or value system). Hence, instead of understanding rights of rivers as *either* being about human rights *or* nature's rights – or to view them as in competition with each other – we would point out that they can also be constructed as interrelated concerns.

Our use of the term *riverine* rights to describe these cases of granting rights to rivers is tied to this reasoning. The concept *riverine* has a relational quality to it. It means *relating to* a river, *dwelling beside* or being *situated on* a river, *resembling* a river or being *formed by* a river (Oxford English Dictionary 2010, The Free Dictionary n.d, Merriam-Webster Dictionary n.d.). Hence, riverine includes something more than the river alone. The cases of this book are not simply about the rivers but also about the people living along or with them – those who are related to the river (which also includes a larger river ecosystem with many different species). Speaking of *riverine* rights, then, covers this relationality and potential of more-than-river rights. In their attempts to secure the wellbeing of people and rivers together, the legal decisions that grant personhood rights to Whanganui, Atrato, Ganga and Yamuna rivers are cases of *riverine rights.*

This concept reflects our analytical understanding of the rights of river cases in question. In the laws and rulings, the term river is used to describe the right-bearing entity, but *riverine* draws attention also to humans and their many concerns in relation to the river.

Why rivers? Why now? Why rights of nature?

Why rivers?

Since the legal decisions of granting rights to rivers in Aotearoa New Zealand, Colombia and India came in 2016/2017, an increasing number of cases worldwide have recognized the rights of rivers or other water bodies. By the end of 2024, about 20 river's rights cases were passed globally, overwhelmingly in the Americas but also with examples from Bangladesh, Australia[4] and Spain. According to the Eco Jurisprudence Monitor, which registers and attempts to keep track of all earth jurisprudence cases worldwide, there are now 100 proposals of giving rights to freshwater bodies and about 20 for marine environments. Other emergent initiatives exist that have not yet been included in this count, such as the Oslofjord in Norway and Lake Vättaren in Sweden. Also in the case of Ecuador, where the constitution of 2008 recognizes the rights of nature as a whole, almost all of the Constitutional Court's rights of nature verdicts involve water (Tanasescu et al. 2024, 2). Hence, there is a growing trend of giving rights to water bodies and especially to rivers. So, what is it about water and rivers that spur these initiatives?

Rivers – flowing water bodies – carve their way through the surface of earth, shape landscapes and enable life in and along them. They are biodiverse ecosystems and important sites for human activities. While rivers have persisted for almost all of earth's existence (Graham et al. 2010), humans are relatively new in the history of the planet. Nevertheless, today all rivers are affected by humans, and rivers shape and have historically played important parts in shaping human societies and lives. Rivers are important sites of connection, connecting places, peoples, materials, beings and diverse bodies of water. They are links between different ecosystems, running from glaciated mountains and highlands, through deserts, forests and lowlands, from lakes and springs to the ocean – connecting sweet water with salt water.

Rivers have historically, and to this day, had the capacity to entice and move people, give rich sensory experiences, and evoke emotions, creativity and imagination. They have an aesthetic and imaginative quality to them. Many rivers have religious and cultural significance and are embedded with meanings. Rivers have a place in many origin myths and are in some instances tied to ideas of cleanness and pollution, purity and impurity. They are often places of recreation and creation of kinship. Given the long-existing relationships between peoples and rivers on all continents and corners of the earth, the quality and type of relationships vary greatly. Rivers can be valued as everything from a resource for extraction, to a lifegiving source or as kin. The same river can have multiple meanings, values and uses for the same or different people living along it – cutting across or overlapping in time and space. There are close relationships of care between rivers and individual human persons and groups.[5]

In the history of humans, rivers have provided food, been important trade routes, irrigated fields and allowed for agriculture along them, enabled concentrated populations and been central in state formation. Rivers have been sites of colonial conquest, scientific investigations and societal development, which since the industrial revolution have accelerated in speed and scope.[6] From the utilization of the power of free-flowing water to travel along, to the first water mills for power production, to small-scale diversion, re-direction and damming of rivers, to yet again heavy alterations of river flows, today two-thirds of the rivers globally are impacted by human infrastructure development (Grill et al. 2019), while climate change, pollution and other human interventions affect all rivers on earth. Hence, no river on the planet is unaffected by humans – directly or indirectly – and often one river is subject to several forms of impacts (Wohl 2019).

Although humans have tried to control the flow of rivers for a long time, the extensiveness and acceleration of human alterations of rivers have recently been of a whole other scale. Also, the many efforts for political and territorial control of rivers and river peoples, and the many conflicting human interests in rivers, make rivers politically contested sites. Hence, there is a close connection between rivers and power – and oppression (of rivers, their beings and human persons). Moreover, the vast societal encroachments on rivers have contributed to ecological and environmental crises – related to poor water quality, reduced flows, biodiversity loss and species extinction. Rivers are sites of violence, health threats, conflict, oppression and disaster. Rivers embody societies' traits and problems, and humans embody rivers.

While almost all humans relate to a river, everyone relates to water, embodies water – nearly two-thirds of the human body consists of water – and we are dependent on water for biophysical survival. Therefore, the importance of access to water and the health of water bodies resonates everywhere. Water is strongly associated with life, and rivers are seen as life-veins. When water bodies are threatened, life is threatened, and the matter becomes intimate in a more universal sense than with other entities in the environment. "Water is life" is a central mantra in environmental movements, and rivers and other water bodies are situated at the heart of the current expansion of rights of nature. Riverine rights take a hold in very different places across the earth, in the context of, on the one hand, extensive and longue durée people-river relationships and the multiple significance of rivers for human society, and, on the other hand, rivers as particular sites of encroachments, environmental disasters and accelerated threats.

Why now?

The larger scientific community is crystal clear about the critical state of our planet, with serious consequences for the future of humanity and earth (IPCC 2021). They warn that there is an urgent need to slow down the intensification of the multiscale environmental crises of our time. The Anthropocene – suggested as a term by the turn of the millennia for the geological epoch following the Holocene but recently rejected by the International Commission on Stratigraphy (Carrington 2024) – is descriptive of the extensive, permanent human induced changes to the earth's geology, ecology and climate (Steffen et al. 2007). It is grounded in imperialism, capitalism and extraction of resources that mean both a suppression of local, often Indigenous people and a colonization of nature. Moreover, the widespread use and interest in the concept of the Anthropocene (including its critique) is tied to changing understandings of the threats the planet's ecosystem is facing (Kelly 2018). This particular moment in time, with a heightened awareness and urgency related to the state of the earth, is also when we are seeing the first cases of granting rights to nature and a turn towards granting personhood rights to water bodies in particular.

In many places, the environmental catastrophes that follow from extractivism and climate change manifest as struggles over water. The redirection of, encroachments on and deterioration of rivers and river ecosystems lead to experiences of water scarcity and poor water quality. Since water is vital for biophysical survival and integral to social life (Krause and Strang 2016, Linton and Budds 2014, Orlove and Caton 2010, Strang 2009), this affects many aspects of people's lives. The extensive negative consequences of environmental transformations highlight the urgent need for better governance of human interaction with the environment. Many call for a paradigm shift that challenges the dominate modern, colonial, anthropocentric ways of treating the environment and is more inclusive of other-than-human entities and perspectives and ontologies that are founded on ideas and practices of a mutual constitution of the human and other-than-human. It is under these circumstances that rights of nature emerge as an alternative that seeks to redefine human-environment relationships, starting with modern law. Inspired by threads of Euro-American environmental philosophy and Indigenous worldviews and ways of being in the world, rights of nature has been sketched out as a solution or the best way forward in a growing, global movement (Boyd 2017).

However, as Tanasescu (2022) points out, the development of rights of nature can also be understood in relation to an increasing expansion of liberal rights. This is a trend of judicialization and depoliticization, where courts come to dominate over political deliberation. The very idea of rights is based on thoughts of individual freedom – an individualism that started out as a project of human liberty and has had far-reaching consequences including outcomes that are unjust and undemocratic. Moreover, the same ideology of liberalism underpins the global dominance of the capitalist economic system, which has enabled the acceleration of human-induced climate change and environmental degradation that rights of nature legislations are trying to combat. These are some of the circumstances, and perhaps paradoxes, of the development of rights of nature.

In Chapter 8, Brandshaug and McNeish go more into detail about the global currents that have opened a space for the expansion of rights of nature at this particular moment in time, such as changes in international institutions and policy, trends in global environmental and rights movements, and developments in natural and social sciences regarding human-environment interconnections. As the chapters of this book will demonstrate, each of the cases of personhood for rivers that we examine has its own history and circumstances that have paved the way for the development of their version of rights of nature or rights of rivers that is particular to that place although part of larger "translocal assemblages" (Kinkaid 2019).

This book contributes to current debates on rights of nature by critically examining the three cases and explores to what degree there is a potential or realistic possibility for rights of nature or rivers to challenge and change dominant paradigms responsible for causing these troubles in the first place.

Why rights of nature? Critical perspectives

As explained above, rights of nature are presented as a paradigm shift, a change from an anthropocentric way of managing relations with the environment (based on the separation of humans and the environment) to an ecocentric one (based on the view that humans are part of the biosphere). In light of the ever-expanding circle of rights-bearing subjects within the history of liberal law, giving rights to nature may be seen as simply a logical continuation of a progressive development process, as Stone (1972) argued. Still, it is a radical step to afford rights and legal personhood to nature, specific ecosystems or a river. The granting of rights and legal personhood to entities which legislating bodies may not see as sentient, or as having intentions or consciousness, is a legal innovation. And to the extent that it implies moving from seeing nature as an object, a resource to be used for the benefit of humans, towards seeking to balance the interests of all – humans as well as other-than-human entities – then it surely is a paradigmatic shift.

However, one type of criticism that has been levelled against the idea of rights of nature claims that in reality the case is the opposite: rights of nature is really just a thin gloss that promises radical change while in practice allows for an underlying and destructive logic of human exploitation of nature and of other humans to persist (Youatt 2017, 40). Variants of this radical critique have been raised from different positions – environmentalist, Marxist, decolonial or Indigenous. Youatt (2017) cites John Livingstone:

> I don't think I want a redwood grove to have rights. Rights are political instruments –legal tools. We hear a lot of talk about 'extending' rights to nature. How bloody patronizing! How patriarchal for that matter. How imperialistic. To extend or bestow or recognize rights to nature would be, in effect, to domesticate all of nature – to subsume it into the human political apparatus.
>
> (Interview with Livingstone in Jensen 2004, 62)

Other critics focus on the rights of nature serving as a veil to allow an economic model based on extractivism to continue. In a study of the effects of Ecuador giving constitutional rights to nature, Arsel concludes that the commercialization of nature and natural resources is allowed to continue unabated (Arsel 2012). From the perspective of Australian Aboriginal people, Marshall questions the idea of rights of nature. She claims the introduction of rights of nature ignores Indigenous traditions of caring for nature and is primarily a way for colonial or Western law to extend its reach and dominance over other understandings and forms of governance (Marshall 2019, see also Viaene 2022).

Tanasescu argues that two trends that characterize the current historical situation – "the intensification of human pressure on the environment and the expansion of liberalism in the guise of increasing numbers, and kinds, of rights"– are closely connected (Tanasescu 2022, 10). The political economy of perpetual expansion of capitalism needs and thrives on the liberal, rights-based order which the modern state applies selectively to favour global capital expansion. While there is nothing inherent in rights that mean they must favour the interests of capital, the crucial issue is how they are implemented. In Tanasescu's view, rights of nature are not primarily about nature. Rather, they are political, and what is important is the power configurations they facilitate (Tanasescu 2022, 10–17). One determining factor in this respect is the issue of who gets to represent nature's interests (Tanasescu 2022, 139).

In our view, these critical perspectives are insightful and necessary to keep in mind when examining our cases. Our approach, however, is very much empirical and inductive. Our primary task is the investigation of the three cases. The different takes on rights of nature – both the positive and the critical – are with us as issues to be explored in each case. But our intention has been to approach them with open minds and let our conclusions reflect what we found in each instance. The following section describes in greater detail how we have gone about our study.

Researching riverine rights

Interdisciplinarity, sociolegal studies and legal pluralism

This book is the outcome of a sociolegal study. The focus is on law – the granting of legal personhood to rivers through legislation and judicial decisions. But the objective of the study is to understand law in its social, cultural and political context. We study "law in action" as opposed to "law on the books" (Menkel-Meadow 2019, 35). Thus, we are not only after the intentions of judges and lawmakers, and how the decisions to give legal personhood to rivers relate to other elements of the law and its system of enforcement. We also wish to understand the broader social processes through which these decisions came about, as well as how they are followed up in practice in each case, and what their effects are.

With this objective, the research project has been interdisciplinary, bringing together legal scholars and social scientists. Among the latter, the project includes many anthropologists, but disciplinary backgrounds also include history, environmental economics, business and organizational studies, political science, and water science, policy and management.

Researchers of the Riverine Rights Project

Project coordination

Axel Borchgrevink, social anthropology, project leader
John Andrew McNeish, social anthropology, deputy project leader
Malene K. Brandshaug, social anthropology

Aotearoa New Zealand

Elizabeth Macpherson, law
Miriama Cribb, business and organizational studies, management
Axel Borchgrevink, social anthropology

Colombia

Catalina Vallejo Piedrahíta, law
John Andrew McNeish, social anthropology
Jenny Moreno Socha, political science
Daniel Alfonso León, history
Arlinton Cuesta Mosquera, law

India

Bibhu Prasad Nayak, environmental economics
Rahul Ranjan, social anthropology/history
Kavita Upadhyay, water science, policy and management

Norway

Malene K. Brandshaug, social anthropology

Epistemological issues make such interdisciplinary work challenging, as different disciplines ask different questions, use different methods, rely on different types of data, relate to different theories and pursue different types of explanations and understandings. In carrying out this study, we have sought to overcome these differences in approaches through frequent contact and communication throughout the more than five years we have been working together. By carrying out data collection in the case studies jointly across disciplinary boundaries, and by regularly partaking in online meetings as well as attending four in-person workshops, bringing the whole team together, we have created a common understanding of the project. Thereby we allow our different perspectives to complement and support each other, rather than being a source of confusion and disagreement.

This is not to say that we have ended up seeing all things in the same way. The disciplinary differences remain, as do our (relatively minor) differences when it comes to certain theoretical stances. In the writing of this book, the authors of the chapters have been allowed to write and conclude as they see fit. The observant reader may find that there are some minor differences in perceptions and assessments between chapters. We see this as enriching the book rather than representing a problem. And we should emphasize that in the main these differences are small.

Given its interdisciplinary character, sociolegal studies is a heterogenous field. Much of it is dominated by legal scholars (Anders 2015). In this study, however, the project leaders, as well as the editors of this book, are anthropologists. This is by no means unique; the anthropology of law is a well-established field within sociolegal studies (McConnachie 2019). One disciplinary difference between legal scholars and social scientists (such as anthropologists) relates to normativity. Within social science, the emphasis tends to be on the descriptive study of the empirical existence of normative systems within a given social setting (Merry 1988, 871). Legal scholars, on the other hand, according to von Benda-Beckmann, are generally concerned with "interpretations of general legal abstractions with respect to concrete problematic situations (cases) and/or philosophical reflections on what and how law should be" (von Benda-Beckmann 2008, 94, see also Anders 2015, 413).

In the present study, our focus is on the "is" and not on the "ought" (while we remain motivated by the curiosity of how rivers could be better protected amid the ongoing ecological crisis). Our primary aim is to understand what is happening in our three cases. When it comes to rights of nature – or legal personhood for rivers – we have no preconceived ideas about the usefulness of such legal approaches. In this respect, we are agnostics, not believers. And as one of our aims is to explore whether there are lessons to be learnt and ways of doing things that could be useful elsewhere in these cases, there is a need to examine them critically.

Anthropology has always been just as interested in the informal ways that social interaction is structured as in the formal ones. Therefore, it is not surprising that the anthropology of law has been attracted to the perspective of legal pluralism – being open to the possibility of the coexistence of multiple legal orders within a particular socio-cultural context. Within sociolegal studies, there is an important dividing line between those who see law as only referring to the law of the state, and those that are open to legal pluralism and "the insight that in most societies there is a plurality of legal orders" (von Benda-Beckmann 2008, 97). This study belongs to the latter category.

Legal pluralism can be understood as an evolving field where different generations can be perceived. The origin of the term is often attributed to studies of colonial or postcolonial societies, where an external, European system of law is imposed on pre-existing systems of norms, conflict resolution and regulation. What Merry in the 1980s termed "the new legal pluralism" was an approach that recognized different legal orders also in modern societies that had not been colonized by expanding the concept of the legal system to also include the non-legal ordering that takes place in institutions such as corporations, factories and universities, as well as the informal establishment and enforcement of norms and rules within families and other social groups (Merry 1988). Some 20 years later, Berman writes about a new "new legal pluralism" being applied to the international and transnational realm. In response to overlapping jurisdictional assertions by states, international bodies and non-state entities, and the accompanying evolving global norms, the legal pluralist perspective is a useful framework for capturing this hybrid field of contesting and partially overlapping normative systems (Berman 2009, 226).

These perspectives are also highly appropriate for the cases that we are studying. They all pertain to postcolonial societies where the legal personhood decisions in one way or another are intended as ways of bridging between formal legal systems inherited from the colonial powers and other and local normative orders. Thus, these cases are explicit examples of the types of legal intermingling that classical legal pluralism has focused on. In addition, rights of nature is also an emerging and growing global legal innovation directly aimed at reforming state and international law. Studying the way this idea evolves and spreads, and how it interacts and merges with local norms, state and non-state legal systems when realized in practice in different parts of the world – such as in the personhood for rivers in our three cases – falls squarely within the third generation of legal pluralist studies that Berman writes about.

Comparing contextualized case studies

We believe there are useful methodological lessons to be drawn from the way this project has been designed and executed. As mentioned, the Riverine Rights project was initiated as a response to the curiosity sparked by the three cases of rivers having been given legal personhood. A case study design, comparing these three cases, was therefore a given from the outset. Moreover, we wanted to understand the cases on the background of a full understanding of their particular social, economic and cultural context. An inductive approach was followed, starting with comprehensive data collection in all countries. The research team has explored legal practices, river practices and governance practices, and conducted document analyses of legal texts, policy documents and regulations, as well as newspaper stories and other narratives. This includes dozens of qualitative interviews, informal conversations and participant observation with groups that are differently positioned, such as local and Indigenous people, representatives from governance agencies and non-governmental organizations, environmental and Indigenous activists, politicians, legal experts and members of river management bodies. Ethnographic methods were particularly important for connecting with local people, who are most directly affected by the legal acts of giving personhood status to these rivers. In order to understand their relations with the rivers and their concerns related to the environment, livelihoods and other urgent issues, spending extended periods of time in the area and including local researchers on the project team were vital.

The research teams for each country case have consisted of people with complementary disciplinary backgrounds and relevant methodological experience – crucially they are from or have long experience of doing research in that country. Particular attention has been paid to securing the approval and cooperation of local organizations and institutions of river dwellers, of respecting their conditions for allowing us access and of seeking to repay the trust and help we have received by somehow being useful in the local context.

In addition to the three case studies from India, Colombia and Aotearoa New Zealand, one of the project's postdoctoral researchers (Malene K. Brandshaug) has researched international dimensions of rights of nature and personhood rights to rivers and studied an additional case of environmental concerns pertaining to water bodies in Norway, including the dawning emergence of rights for rivers and fjords there. Although the case from Norway is not included in this book, insights from this research have contributed to the overall understanding of local people's relationships to their waters, mobilization for protection of water bodies and the development of rights of nature more broadly.

Aotearoa New Zealand

By Miriama Cribb

Given the Whanganui River case's intertwinement with Indigenous Māori concerns, the research in Aotearoa New Zealand was carried out with Indigenous research methodologies in mind. Indigenous methodology is rooted in relational connections – with and between people, the environment and places, with the cosmos and ideas – and through them ultimately raising one's consciousness and insights into Indigenous worldviews (Wilson 2008). In Aotearoa New Zealand, *kaupapa* Māori methodology can be understood within the broader context of Indigenous methodology because it is related to being Māori; it is connected to Māori philosophy and principles, the validity and legitimacy of Māori, the importance of Māori language and culture, and is concerned with the struggle for autonomy over Māori cultural wellbeing (Smith 1997). *Kaupapa* Māori methodology for this project has implied doing research with and about the Whanganui River and Whanganui *iwi* in a relational, reciprocal and respectful manner.

Given the importance of relationality for Whanganui *iwi*, our research project should not only seek to achieve our own aims and objectives. There should be reciprocity and benefits for Whanganui *iwi*. *How* we do research is just as important as *what* we research. In addition to obtaining ethics clearances at our respective universities, our research methods began with engagement with the Whanganui *iwi* via Ngā Tāngata Tiaki, the post-settlement governance entity of the *iwi* for the Whanganui River. This included a long online interview that project leader Axel Borchgrevink and Aotearoa New Zealand lead researcher Elizabeth Macpherson had with former chairperson Gerrard Albert, where both sides got to know and learn about the other. Through this engagement, a position was made available for a local Māori researcher (Miriama Cribb) to join the project team. We were to familiarize ourselves and uphold the interim research protocols outlined by Ngā Tāngata Tiaki before, during and after the research project.

In carrying out the study, we have attempted to the best of our understanding to adhere to the appropriate protocols and practices of Whanganui *iwi*. Cribb has lived in Whanganui her whole life, Borchgrevink spent three months in the region and Macpherson visited regularly throughout the project period. Relationship-building was at the forefront of our minds, and we attempted to engage and meet with as many local Māori and non-Māori as possible. Seeking to understand and experience the culture and history of the Whanganui River, we interviewed and socialized with key stakeholders involved in Te Awa Tupua, met with local people both Māori and non-Māori, visited sites of significance, *marae* (villages, local gathering places), took part in meetings and events related to the river, and travelled along and spent time on the river.

We have supported building the capacity of local researchers and connected them to other researchers, projects and international networks. We have sought local and tribal advice during dissemination; for example, through fact checking during the development of publications. Where appropriate, we have been conscious of and considered the role of research in supporting the health and wellbeing of Te Awa Tupua. Given the varying disciplines, expertise and broad backgrounds of our research team, we committed to clarifying our (research) positionalities while trying to uphold and maintain everyone's dignity.

Colombia

By Catalina Vallejo Piedrahíta

The research in Colombia was initiated by Catalina Vallejo Piedrahíta and John Andrew McNeish, who began with virtual interviews in 2020. These interviews included key judicial actors from the Constitutional Court of Colombia and other lawyers, who shared their perspectives on the rights of nature and on the Atrato River decision and its broader context. In 2021, the team carried out a field visit to Quibdó, where they navigated a section of the Atrato River by boat, observing the impacts of gold mining and extractivism, state presence and absence, local cultural practices and the strong relationship between communities and the river. This visit also marked the beginning of their relationship with the river guardians.

Subsequently, political scientist Jenny Morena Socha, experienced in working with social movements for river protection in Colombia, joined the team, along with Arlinton Cuesta Mosquera, a lawyer from the Chocó region, *atrateño*, and a Master's student at UNAULA University, the project's partner institution in Colombia. Later, historian and audiovisual expert Daniel León Alfonso also joined the team. The full research team conducted in-depth interviews in Quibdó and Bogotá in 2022 and 2023, engaging with state and community stakeholders, including members of the Embera Dobidá people and the *consejos comunitarios*, the autonomous authorities of Black communities. A visit to Quibdó by researchers from Aotearoa New Zealand and India allowed meetings and exchange between Māori representatives and the Atrato River guardians, which included a symbolic delivery of gifts from the Māori community to the Atrato guardians, as well as dialogue on their shared experiences in river defence, sovereignty and the common goal of collaborative governance with state actors.

As part of their commitment to the community, the research team produced videos to support the river guardians in their educational efforts about the ruling and its implementation. They also prioritized establishing respectful relationships, sharing findings with participants and ensuring accessibility by translating research findings into Spanish and publishing an article in a Colombian academic journal.

Additionally, during their visits, the team was invited by river guardians to enjoy trips and recreational activities in other clean, vibrant tributaries of the Atrato. This experience was a delightful surprise, as they had initially assumed that pollution and violence dominated the river's narrative. Instead, they witnessed the river's stunning natural beauty, teeming with life, and experienced firsthand the deep emotional and cultural connection between local communities and their waterways. These moments offered a glimpse into the joy, resilience and strong bonds that people in the Atrato basin share with their rivers.

India

Data collection in India posed different challenges than the other cases. On the one hand, the immense scale of the Ganga and Yamuna river basins, with their multitudes of people and enormous cultural and ecological diversity, meant that it was impossible for the project to get a representative overview through anthropological methods. And on the other hand, as the rulings by Uttarakhand High Court that declared the Ganga and Yamuna rivers to be legal persons were quickly stayed by the Supreme Court of India, they have had little practical effect. There is no implementation of the decisions, and there are therefore not any direct outcomes for us to document and analyse.

Research in India has focused on two main issues: tracking the legal process that led to the rulings and exploring selected cases that illuminate the way in which people relate to the rivers. Most of the research has been carried out in Uttarakhand. Lead researcher Bibhu Prasad Nayak visited the state repeatedly over the project period. He and Kavita Upadhyay (a journalist native to the state with extensive experience of writing about environmental issues) have explored the court cases, from the claims raised to the rulings. Moreover, they have investigated the 2017 decision by the Madhya Pradesh State Assembly to declare Narmada a living entity with rights. Nayak has also done research in legal circles at Central level, including the National Green Tribunal, on attitudes to rights of nature in the Indian context.

Rahul Ranjan did six months of fieldwork in Uttarakhand, mainly studying how people relate to and engage with the sacred rivers of Ganga and Yamuna. He studied how flood disasters – natural as well as induced by infrastructural development projects – have affected people, livelihoods and perceptions, and popular movements against dams and road constructions. He has also done research among a group of Hindu ascetics engaged in the struggle for cleaning the Ganga River, including by hunger strikes that have led to the deaths of two of them.

Comparison

The in-depth data collection on the different cases is the foundation of the project. These case studies have yielded valuable results on their own that have been documented in a series of publications (see for instance, Cribb et al. 2024, Cuesta Mosquera and Vallejo Piedrahíta 2024, McNeish and Moreno Socha 2024, Ranjan 2024, Upadhyay and Nayak 2024). The rich empirical data material and the comprehensive insights into the dynamics of each case are also the basis for the next step: juxtaposing and comparing the cases to see what we can learn from them. What are the similarities? What are the contrasts and differences between them? What types of social mechanisms can be found that allow us to draw more general conclusions from our study?

In order to make the study into something more than three free-standing case studies, we have worked in ways designed to ensure their compatibility and comparability, and we have organized the project to ensure the sharing of findings between case studies and the researchers working on them. When the project started, in 2020, the pandemic prevented us from realizing our field-work according to the schedule we had planned. It took more than a year before data collection was fully underway. However, this was in unexpected ways a blessing. The funder of the project, the Norwegian Research Council, allowed us to extend the project period for an additional year, until the end of 2024. This permitted us to dedicate the first project year to developing a common understanding of the issues and an approach to be applied in all case studies. Monthly online meetings, with prepared interventions and sometimes invited guests, allowed us to get to know each other and develop team spirit and commitment to our joint undertaking.

As the project advanced, with fieldwork taking place in the case countries, we also strengthened exchange between these research components. In April 2022, the whole team met for the first time for a workshop in India, with seminars with legal scholars in Delhi and visits to different parts of Uttarak-hand and meetings with activists, scholars and people living along the Ganga. In January 2023, the team met up again in Aotearoa New Zealand for a workshop that started at the University of Canterbury in Christchurch and thereafter took us all on a trip along the Whanganui River from its sources to the sea, including one stretch in canoes, with visits to communities and meetings with representatives and organizations along the way. In December 2023, we held a workshop in Colombia for the whole team, visiting the Atrato River and the regional capital of Quibdó, as well as allocating time for seminars with researchers in Medellin. And in August 2024, the project held its closing conference in Oslo. This allowed us a final meeting of the team for comparing our findings, discussing conclusions of the project and commencing the work with this book, before two days of presenting our research and engaging in conversations with other academics and the general public.

This way of working has allowed all involved researchers to have visited and learnt about the other cases, in addition to the in-depth knowledge they each developed on the case they had been actively researching. This has had twin benefits, both in terms of being able to draw comparisons across cases and see new aspects of the case one had been working with. This book is about the insights we have developed through this way of working. Although not all project members are co-authors of this book, they have all contributed to the research on which this book is based and have offered their comments on chapter drafts.

Overview of the book

After this introductory chapter, there are chapters on each of the three cases of rivers with legal personhood we have been studying. In Chapter 2, Miriama Cribb, a born and bred Whanganui Māori and researcher, writes about the Te Awa Tupua (Whanganui River Settlement) Act from the perspective of an insider and researcher. Her chapter describes the background for the settlement between the Crown and the Whanganui *iwi* and emphasizes how the integration of *kawa* – Māori law or values – into the law makes this a far-reaching example of legal pluralism. The Colombia chapter, Chapter 3, is co-authored by the researchers involved in that case study: Catalina Vallejo Piedrahíta as first author, with John Andrew McNeish, Jenny Moreno Socha, Daniel Alfonso León and Arlinton Cuesta Mosquera. It details the historical background of the Chocó Department, with continuities in gold extraction, armed struggles and violence and marginalization of Indigenous people and afro-descendants from colonization to this day. The chapter shows how the ruling of the Constitutional Court that gave biocultural rights to the Atrato River and its people was an attempt at addressing the urgent problems this ongoing history has created for the people living along the river, and it further examines the reasons for the limited success in resolving these problems. Next, in Chapter 4, Axel Borchgrevink and Kavita Upadhyay write about the Uttarakhand High Court rulings that gave legal personhood to the Ganga and Yamuna rivers, as well as to the glaciers and surrounding areas from which these rivers flow. They give a detailed examination of the court cases leading up to these decisions, as well as a critical analysis of the judgments.

The following chapters provide comparative perspectives on these three cases. In Chapter 5, Malene K. Brandshaug, with Miriama Cribb, Jenny Moreno Socha, Daniel Alfonso León and Arlinton Cuesta Mosquera, write about people's relations to and understandings of the rivers they live alongside in the three countries, and discuss how these are reflected in the legal documents. Blending academic writing with extensive quotes from people of the rivers and excerpts from origin stories, novels and poetry, as well as photographs, the intention is to convey the many ways in which people's relations to their surroundings transcend the instrumental view of nature as simply resources for human consumption. In Chapter 6, Catalina Vallejo Piedrahíta, Nataly Montoya-Restrepo, Camilo Arango Duque and Julián Gaviria Mira examine the cases from the perspective of legal scholarship. They discuss the possible philosophical justifications for introducing the rights of nature, examine potential pitfalls and compare the cases according to key legal dimensions. They conclude that while these cases offer valuable insights into the transformative power of the rights of nature framework, they also underscore the need for careful consideration of local ecosystems, cultural values and governance structures. In Chapter 7, Axel Borchgrevink compares and discusses the political dimensions of these cases. There are significant differences between them, and the chapter examines colonial histories, differences in state capacity, the political economy of natural resource extraction and contestations over knowledge and values. The chapter demonstrates how these dimensions help understand outcomes in each case.

The three cases of legal personhood for rivers are all in some ways examples of rights of nature, and their occurrence is somehow related to the evolving global ideas of nature's rights – even if our case studies show that this relationship is by no means direct and straightforward. Chapter 8, by Malene K. Brandshaug and John Andrew McNeish, explores issues related to the question of why these three groundbreaking legal decisions of personhood for rivers appeared almost simultaneously in three so different countries. They point to global conjunctures and developments within natural and social sciences – and a broad constellation of movements, organizations and ideas – that together go some way towards explaining the current growing interest in rights of nature. They reflect on some of the opportunities and potential pitfalls of granting rights to nature.

The final chapter, Chapter 9, seeks to pull the threads from the book together by returning to the three questions that sparked the project: Why was legal personhood granted to rivers in these cases? What have been the outcomes of these decisions? What can we learn from them?

Notes

1 Public Interest Litigation (PIL) can be defined as legal action undertaken by lawyers representing individuals and communities with the explicit aim of altering certain aspects of the political, social or economic status quo. This form of litigation assumes that existing social conditions hinder the participation and benefits of certain subordinated groups, and it primarily seeks to amplify their voices and interests (Loewy 1999, 1869–1870 as cited by Ucín (2017, 249).
2 https://ecojurisprudence.org/dashboard/?map-style=political.
3 Apparently, Justice Black dissented the granting of personhood to railroad corporations with the argument that corporations had "neither race nor color" (Stone 1972, 465, note 51).
4 https://riverresourcehub.org/resources/rights-of-rivers-global-map/.
5 For an exploration of different cultural forms of engagement with water, see Strang (2009).
6 For the history of the Nile, see Tvedt (2012), and for the role of rivers/water in state formation, see Scott (2017).

References

Anders, Gerhard. 2015. "Law at its limits: Interdisciplinarity between law and anthropology." *The Journal of Legal Pluralism and Unofficial Law* 47 (3): 411–422.

Arsel, Murat. 2012. "Between 'Marx and markets'? The state, the 'left turn' and nature in Ecuador." *Tijdschrift voor economische en sociale geografie* 103 (2): 150–163.

Berman, Paul Schiff. 2009. "The new legal pluralism." *Annual Review of Law and Social Science* 5 (1): 225–242.

Berry, Thomas. 2010. *Evening thoughts: Reflecting on earth as a sacred community.* Catapult.

Boyd, David R. 2017. *The rights of nature: A legal revolution that could save the world.* ECW press.

Carrington, Damian. 2024. "Geologists reject declaration of Anthropocene epoch." *The Guardian*, March 22. https://www.theguardian.com/science/2024/mar/22/geologists-reject-declaration-of-anthropocene-epoch.

Cribb, Miriama, Elizabeth Macpherson and Axel Borchgrevink. 2024. "Beyond legal personhood for the Whanganui River: Collaboration and pluralism in implementing the Te Awa Tupua Act." *The International Journal of Human Rights*: 1–24.

Cuesta Mosquera, Arlinton, and Catalina Vallejo Piedrahíta. 2024. "Take me to the River: Have riverine rights enhanced community participation in environmental governance within the Atrato River basin?." *The International Journal of Human Rights*: 1–19.

Cullinan, Cormac. 2011. *Wild law: A manifesto for earth justice.* Bloomsbury Publishing.

García Ruales, Jenny, Katarina Hovden, Helen Kopnina, Colin D. Robertson and Hendrik Schoukens, eds. 2024. *Rights of nature in Europe: Encounters and visions.* Taylor & Francis.

GARN (Global Alliance for the rights of nature). 2022. *Constitution of the Global Alliance of the Rights of Nature.* https://www.garn.org/wp-content/uploads/2024/02/FINAL_GARN-Constitution_English.pdf.

Graham, Steve, Claire Parkinson and Mous Chahine. 2010. "The water cycle." Accessed 20 September 2024. https://earthobservatory.nasa.gov/features/Water.

Grill, Guenther, Bernhard Lehner, Michele Thieme et al. 2019. "Mapping the world's free-flowing rivers." *Nature* 569: 215–221. doi:10.1038/s41586-019-1111-9.

IPCC. 2021. *Climate change 2021: The physical science basis.* Contribution of Working Group I to the Sixth Assessment Report of the Intergovernmental Panel on Climate Change. Edited by Valerie Masson-Delmotte, Panmao Zhai, Anna Pirani et al. Cambridge University Press. doi:10.1017/9781009157896.

Jensen, Derrick. 2004. *Listening to the land: Conversations about nature, culture and eros.* Chelsea Green Publishing.

Kauffman, Craig M., and Pamela L. Martin. 2021. *The politics of rights of nature: Strategies for building a more sustainable future.* MIT Press.

Kelly, Jason M. 2018. "Anthropocenes: A fractured picture." In *Rivers of the Anthropocene*, edited by Jason M. Kelly, Philip Scarpino, Helen Berry, James Syvitski and Michel Meybeck. University of California Press.

Kinkaid, Eden. 2019. "'Rights of nature' in translation: Assemblage geographies, boundary objects, and translocal social movements." *Transactions of the Institute of British Geographers* 44 (3): 555–570.

Krause, Franz, and Veronica Strang. 2016. "Thinking relationships through water." *Society & Natural Resources* 29 (6): 633–638. doi:10.1080/08941920.2016.1151714.

Linton, Jamie, and Jessica Budds. 2014. "The hydrosocial cycle: Defining and mobilizing a relational-dialectic approach to water." *Geoforum* 57: 170.

Macpherson, Elizabeth Jane. 2019. *Indigenous water rights in law and regulation.* Cambridge University Press.

Marshall, Virginia. 2019. "Removing the veil from the 'Rights of Nature': The dichotomy between First Nations customary rights and environmental legal personhood." *Australian Feminist Law Journal* 45 (2): 233–248.

McConnachie, Kirsten. 2019. "Law and anthropology." In *Routledge handbook of socio-legal theory and methods*, edited by Naomi Creutzfeldt, Marc Mason and Kirsten McConnachie. Routledge.

McNeish, John-Andrew, and Jenny Moreno Socha. 2024. "Muddy waters: On the problematic political ecology of the Atrato ruling, Colombia." *The International Journal of Human Rights*: 1–21.

Menkel-Meadow, Carrie. 2019. "Uses and abuses of socio-legal studies." *Routledge handbook of socio-legal theory and methods*, edited by Naomi Creutzfeldt, Marc Mason and Kirsten McConnachie. Routledge.

Merriam-Webster Dictionary. n.d. "Riverine." Accessed April 4, 2025. https://www.merriam-webster.com/dictionary/riverine.

Merry, Sally Engle. 1988. "Legal pluralism." *Law & Society Review* 22 (5): 869–896.

Naess, Arne. 1989. *Ecology, community and lifestyle: Outline of an ecosophy.* Cambridge University Press.

O'Donnell, Erin. 2018. *Legal rights for rivers: Competition, collaboration and water governance.* Routledge.

Orlove, Ben, and Steven Caton. 2010. "Water sustainability; anthropological approaches and prospects." *Annual Review of Anthropology* 39: 401–415. doi:10.1146/annurev.anthro.012809.105045.

Oxford English Dictionary. 2010. "Riverine." Accessed 4 April 2025. https://www.oed.com/dictionary/riverine_adj?tl=true&tab=factsheet.

Ranjan, Rahul. 2024. "Sadhu activism: Rights of rivers and facets of religious environmentalism in the Himalayas." *The International Journal of Human Rights*: 1–21.

Scott, James C. 2017. *Against the grain: A deep history of the earliest states.* Yale University Press.

Smith, Graham H. 1997. "Development of Kaupapa Maori theory and praxis." PhD thesis, Auckland University.

Steffen, Will, Paul J. Crutzen and John R. McNeill. 2007. "The Anthropocene: Are humans now overwhelming the great forces of nature." *AM-BIO: A Journal of the Human Environment* 36 (8): 614–621. doi:10.1579/0044-7447(2007)36[614:TAAHNO]2.0.CO;2.

Stone, Christopher. 1972. "Should trees have standing: Toward legal rights for natural objects." *California Law Review*, 450–493.

Strang, Veronica. 2009. *Gardening the world: Agency, identity and the ownership of water.* Berghahn Books.

Stutzin, Godofredo. 1984. "Un imperativo ecológico: reconocer los derechos de la naturaleza." *Ambiente y Desarrollo* 1 (1): 97–114.

Tanasescu, Mihnea. 2022. *Understanding the rights of nature: A critical introduction.* Transcript Verlag.

Tanasescu, Mihnea, Elizabeth Macpherson, David Jefferson and Julia Torres Ventura. 2024. "Rights of nature and rivers in Ecuador's Constitutional Court." *The International Journal of Human Rights*: 1–23.

The Free Dictionary. n.d. "Riverine." Accessed 4 April 2025. https://www.thefreedictionary.com/Riverine.

Tvedt, Terje. 2012. *Nilen – historiens elv.* Aschehoug.

Ucín, María Carlota. 2017. "Litigio de interés público." *EUNOMÍA. Revista En Cultura De La Legalidad* (12): 246–255.

Upadhyay, Kavita, and Bibhu Prasad Nayak. 2024. "Tracing the legal journey of petitions in the Uttarakhand High Court that became springboards for rights of rivers and nature in India." *The International Journal of Human Rights*: 1–20.

Vega Cárdenas, Yenny, and Daniel Turp. 2023. *A legal personality for the St. Lawrence River and other rivers of the world.* Editions JFD.

Viaene, Lieselotte. 2022. "Can rights of nature save us from the Anthropocene catastrophe? Some critical reflections from the field." *Asian Journal of Law and Society* 9 (2): 187–206.

von Benda-Beckmann, Franz. 2008. "Riding or killing the centaur? Reflections on the identities of legal anthropology." *International Journal of Law in Context* 4 (2): 85–110.

Wilson, Shawn. 2008. *Research is ceremony: Indigenous research methods.* Fernwood Publishing.

Wohl, Ellen. 2019. "Forgotten legacies: Understanding human influences on rivers." *EOS.* Accessed 20 September 2024. https://eos.org/features/forgotten-legacies-understanding-human-influences-on-rivers.

Youatt, Rafi. 2017. "Personhood and the rights of nature: The new subjects of contemporary earth politics." *International Political Sociology* 11 (1): 39–54.

2 Te Awa Tupua and rights for Whanganui River, Aotearoa New Zealand

Miriama Cribb

Introduction

What is written about Te Awa Tupua (Whanganui River Claims Settlement) Act 2017, especially in academia, seldom comes from the people of the Whanganui River (that is, tribal members and descendants of the Whanganui River and local communities). There is a large focus on the innovative granting of legal personhood to a river. Te Awa Tupua Act is heavily imbued with and centres on Indigenous worldviews and knowledge (Bargh 2020), recognizing the river in Western law for the living entity Whanganui *iwi* (tribes) have always known it to be (Zartner 2021). This recognition is not merely another cultural viewpoint of the same world, but a broad conceptual outline of a completely different worldview, bringing "informal" legal traditions to a more autonomous interaction with the settler state (Tanasescu 2022). A full appreciation and understanding of the context, content, intentions and purposes of the Te Awa Tupua Act, therefore, requires a layered endeavour (Kovach 2009) encompassing the historical, political, social and especially cultural knowledge underpinning it. This chapter was written by a tribal descendant while completing a PhD on Te Awa Tupua. She was born and raised along the Whanganui River and continues to be active in tribal governance and community affairs.

Te Awa Tupua is the name for the Whanganui River and all its physical and metaphysical elements, from its mountain source to the sea. This chapter tries to understand the story of Te Awa Tupua by examining it in its local context, supported by empirical research, to identify the rationale, outcomes and learnings, against the backdrop of riverine rights and rights of nature. There is a particular focus on the history of how colonization led to the encroachment of sovereignty over the river for Whanganui *iwi*, as well as the Treaty of Waitangi settlement process Te Awa Tupua came through. The Te Awa Tupua Act is a culmination of years of efforts from Whanganui *iwi* to correct the wrongs inflicted by colonialism through government policies and regain what was lost to them in 1840, in particular their rights and responsibilities to the river (Zartner 2021). We provide examples of place-based implementation since the passing of the Act in 2017, illustrating that the story

DOI: 10.4324/9781003581208-2

Figure 2.1 The Whanganui River. Map data ©2025 Google. Elaborated by Daniel Alfonso León.

of Te Awa Tupua goes beyond legal personhood to other equally important aspects, such as the jurisdiction of *kawa* or Indigenous law. Such learnings contribute to a growing nuance of theoretical and practical understandings of riverine rights and rights of nature, which must be equally understood alongside the potential of Te Awa Tupua to improve the overall health and wellbeing of our rivers.

The Whanganui River

The Whanganui River is located on the west coast of the North Island of Aotearoa New Zealand and is the most navigable and third longest river in the country, stretching approximately 290 km. The source of the river begins on the northern slopes of Tongariro *maunga* (mountain), which sits amongst and gathers water from a cluster of mountains in the centre of the North Island known as *Te Kāhui Maunga*. The river flows north-west to the township of Taumarunui, then south to the river mouth near Whanganui city, out to the Tasman Sea. The Whanganui River is a prominent feature of the region, covering the boundaries of three district councils and one regional council. Several *iwi* are connected to Te Awa Tupua, with many *hapū* (subtribes) and *marae* (villages, gathering places) located all along the river. The river traverses over 200 km of countryside, through the Whanganui National Park of rich, native lowland forest and untouched wilderness. Through colonization, specifically the Whanganui purchase in the mid-1800s by the New Zealand Company, settlers and their descendants took over large tracts of land along the river and in the Whanganui township. Massive in-migration accompanied the dispossession of *iwi* land, and since the end of the 19th century, settler descendants have made up most of the population in the region and in the country as a whole.

For descendants of Te Awa Tupua, the river is a source of sustenance, a place for gathering food, traditional medicine for healing, and for swimming and bathing, and was the original highway for travel (Takiari 2002). Before any consideration, Te Awa Tupua is part of our genealogical constructs, in that it binds the people to the rivers, mountain, lakes, forests and seas (Turia 2001). Whanganui River women speak about their relationship with their river, describing its moods and beauty and how it is inextricably linked to their employment, their influences, their relationships, their reciprocity, their intergenerational relationships and ultimately their *whakapapa* (genealogical constructs) (Takiari 2002). Traditional insights into Te Awa Tupua, its people and environs are described in song, proverbs and Māori newspapers and manuscripts, showing themes of politics, usage, land, leadership and warfare and speaking directly to the histories, traditions, health and wellbeing of the Whanganui River area, as spoken by ancestors of Whanganui (Tinirau et al. 2020).

Tribal descendants hold the same views and practise the same values for the river as they did many years ago, but today, physically, the river has changed. Developments and impacts such as the diversion of the river for power generation, clearing of eel weirs, introduction of poisons and pests, erosion and land movement, water level changes and changes in weather patterns have seriously impacted the health and wellbeing of the river. Given the physical changes to the river, there are compounding impacts, including the changes in the use of the river and recreational activities and the impact on fisheries, food sources, sustainability and, ultimately, livelihoods. These compounding issues then contribute to the vanishing of *tohu* or various signs in nature used by the old people, impacting their identity and their ability to look after the river, and leading to the disappearance of rituals associated to activities and observation of customary practices (Tinirau et al. 2020). The tribal proverb *ko au te awa ko te awa ko au*, meaning "I am the river and the river is me", illustrates the inextricable link between the river and the people and that when the river is not well, neither are the people.

Te Tiriti o Waitangi and the Treaty of Waitangi

According to Anderson, Binney and Harris (2014), 800 years or so ago, Māori migrated from various parts of Polynesia to a place called Aotearoa. Europeans arrived some 500–600 years later to a country they called New Zealand – a name Māori associated with English participation in the country (Mutu 2004). Mutu describes the issues and role of mistranslation between *Te Tiriti o Waitangi*, a Māori language treaty signed by over 500 Māori chiefs and the Queen of England, and an English document known as the Treaty of Waitangi. In 1840, *Te Tiriti o Waitangi* and the Treaty of Waitangi were signed in good faith between the British Crown and *rangatira* (chiefs) (Hayward and Wheen 2004). In short, *Te Tiriti o Waitangi* allowed the Queen of England to take responsibility by assuming a role of *kāwanatanga* (described as governance) for English immigrants while at the same time agreeing to and supporting the authority (described as *tino rangatiratanga*) of chiefs and subtribes over their lands, homes and everything they valued. *Te Tiriti o Waitangi* also agreed that as part of her *kāwanatanga* role, the Queen would protect Māori and afford them all the rights of British citizens. The Treaty on the other hand conveys a different message, whereby *rangatira* ceded their sovereignty (called *mana* in the English document) and government to the Queen of England in exchange for certain protections (Waitangi Tribunal 2014). The term treaty in this chapter refers to both texts together and or to the event without specifying either document.

Since 1840, there have been ongoing discussions and analysis about the exact nature of the agreement that was reached in both *Te Tiriti o Waitangi* and the Treaty of Waitangi. A lot of the contention has been over how the Crown, represented by the New Zealand government, have used the treaty to further their interests and to validate certain assumed rights (Orange 2015). From the early 1900s, Māori engaged in negotiations with the Crown in an attempt to settle Māori grievances against Crown breaches of the treaty (Wheen and Hayward 2012). In 1975, the Treaty of Waitangi Act was passed, establishing the Waitangi Tribunal, whose role is to investigate Crown breaches of the treaty, which have resulted in prejudicial treatment and loss for Māori (Waitangi Tribunal 2020). Crown breaches against the principles of the Treaty of Waitangi have resulted in over 3,000 claims lodged with the Waitangi Tribunal from Māori seeking the return of stolen and alienated lands, waters and other resources, protection of the natural environment, recognition and restoration of language and culture, equitable access to commercial opportunities and government resources and the upholding and recognition of Māori sovereignty (Mutu 2018).

Through the Treaty of Waitangi Act, the Waitangi Tribunal was established as a standing commission of inquiry. A claim to the Waitangi Tribunal usually follows four steps – pre-negotiation, negotiation, ratification and implementation (Andrew 2008). Claimant groups, often represented by *iwi*, can engage in either direct negotiations with the tribunal or proceed through a hearing process whereby the Waitangi Tribunal hears evidence from claimants to substantiate their claim. The tribunal is made up of both Māori and non-Māori members, appointed for their expertise in particular areas. The negotiation of a claim culminates in a deed of settlement, which signals the resolution of prescribed historical grievances between a Māori claimant group and the Crown (Crocker 2014). A deed of settlement will usually consist of an agreed historical account, a Crown acknowledgement and apology, as well as cultural, financial and commercial redress. A settlement is considered full and final in that it cannot be relitigated by a claimant group; however, a treaty settlement can never be full and final, in that tribes could never regain or be fully compensated for what was lost at the signing of the treaty. More importantly, Māori tribes will continue to challenge the Crown on contemporary and any ongoing injustices and breaches of the treaty.

Political activism for Māori rights in Aotearoa New Zealand has been present since the signing of the treaty, from felling flagpoles to peaceful resistance, from protests and land marches to legal action and the development of political groups. Māori have a long history of fighting for their rights, including land recuperation and language revitalization to redressing treaty grievances and racism in Aotearoa New Zealand. As a result, pan-tribal movements were created, such as the *Kīngitanga* (the Māori king movement) and *Kotahitanga* (Māori parliament movement), as well as local political movements. For Whanganui *iwi*, this includes Pakaitore, which became the site of reasserting tribal ownership, central to Māori political action and cultural revitalization (Pākaitore Historic Reserve Board 2020). Pakaitore, originally a temporary and seasonal fishing village, also used as a trading post, later became the first landing site for *Pākehā* (non-Māori) settlers within Whanganui. In 1995, Māori occupied Pakaitore for 79 days to protest the non-settlement of their treaty claims, contributing to the formation of the Te Awa Tupua Act. Every year on February 28, Pakaitore is remembered for the reclamation and celebration of *Whanganuitanga* – an expression of being a descendant of Whanganui.

The Whanganui River claims settlement

The history of the Whanganui River treaty settlement stretches back to at least 1873, when the first river claim emerged in petitions to parliament regarding timber flotation and the destruction of eel weirs by river steamers (Finlayson and Christmas 2021). Numerous further petitions and legal proceedings followed over the next 150 years, pursued through virtually every court in the country, regarding ownership of the riverbed and the effects of transportation, harbour works, waste disposal, flood control and tourism. In 1877, Whanganui *iwi* objected to harbour board regulations threatening ancient fishing grounds, and in 1895, Whanganui *iwi* took a claim to the Supreme Court over customary fishing rights (Ngā Tāngata Tiaki o Whanganui 2020a). In 1920, settlers proposed a system of hydroelectric dams on the Whanganui River, and in 1959, *iwi* leader Hikaia Amohia formally objected to the headwaters of the Whanganui River being diverted. From 1931, Whanganui *iwi* fundraised for the legal battle to protect their customary rights, and in 1936, *iwi* leaders lodged a petition to challenge the ownership of the river (Ngā Tāngata Tiaki o Whanganui 2020a). These are only some of the events and legal avenues pursued by Whanganui *iwi* in relationship to their river, which were continuously ignored or denied.

Through the trials and tribulations, Whanganui *iwi* consistently maintained that they possess and exercise rights and responsibilities in relation to the Whanganui River in accordance with their protocols and practices. These rights and interests have never been relinquished knowingly or willingly, as the Waitangi Tribunal Whanganui River Report (Waitangi Tribunal 1999) concludes. Whanganui *iwi* were unsuccessful in their legal endeavours to exercise sovereignty and essentially became bystanders of their own river as it was mismanaged with poor decision making and altered in some places, beyond recognition. One of the most significant alterations was the development of the Tongariro Power Scheme, established in the 1960s with no reference to the *iwi*, which has been the subject of litigation ever since (Finlayson and Christmas 2021, 173). This lack of respect for the river had a profound effect on the outlook of many Whanganui descendants, whose association with the river had either been severed or spoiled by new developments. This sense of loss was referred to by witness after witness as part of the Waitangi Tribunal hearings, which are documented in the Whanganui River Report (Waitangi Tribunal 1999).

The Whanganui River Report documents the Whanganui River claim (also known as Wai 167) that was filed by Hikaia Amohia and the trustees of the Whanganui River Māori Trust Board on October 14, 1990. The claim in respect of the Whanganui River was pursued on an inclusive basis for the benefit of "*Te Iwi o Whanganui*", that is, all those groups from the mountains to the sea who affiliate with the river. The Whanganui River report found that in 1840 the Whanganui River was owned in its entirety by the people of the river, namely Whanganui *iwi* (Waitangi Tribunal 1999). The 300-plus page report also found that the Crown, through respective acts, policies and legislation, deprived Whanganui *iwi* of their possessions and failed to protect their *rangatiratanga* in and over their river, which is contrary to the principles set out in the treaty. Because of this, Whanganui *iwi* have been and continue to be prejudiced. It is worth noting that one member of the tribunal, John Kneebone, was not able to support any proposal of Whanganui *iwi* owning natural water or be designated a consent authority in respect of the river under the Resource Management Act (RMA) 1991.

The opposition by Kneebone highlights issues of water ownership; the tribunal accepted that the treaty guarantees to Māori all the properties they in

Figure 2.2 View of the Whanganui River from the hilltop summit Aramoana (also known as Gentle Annie). Photo by Axel Borchgrevink.

fact possessed, of which the river is one part. This was also approved between 1938 to 1962 in a marathon legal case that passed through the Native Land Court, the Native Appellate Court, the Supreme Court, the Court of Appeal and a royal commission of inquiry, until finally being overturned by a new decision in the Court of Appeal. The only difference in this case was that the finding was made with respect to the riverbed because the proceedings started in the Native Land Court, which had jurisdiction only in respect of the land. Thus, the case illustrates the particularity in English law whereby river ownership is perceived in terms of riverbeds or as land with water on it, as opposed to seeing a river as a single entity. The Waitangi Tribunal, in contrast, concluded that what Whanganui *iwi* own is a river, or a particular form of water resource, not a dry bed, and that full and exclusive use of the river is as a water resource (Waitangi Tribunal 1999, xv). The report also illustrates that the position of Whanganui tribal leaders shows they have no objection to the reasonable use of the river by the public, provided that the *mana* or authority of Whanganui *iwi* is respected.

While the story of Te Awa Tupua is unique to the context, history and people of the Whanganui River, the lack of effective management of rivers and natural resources is experienced by other tribes of Aotearoa New Zealand and Indigenous peoples globally. For Māori, resource or environmental management is first guided by the belief in the interrelationships of all parts of the environment (Harmsworth and Awatere 2013), where people are an integral part of the ecosystem. What this means for water governance and management, and what might be more simply summarized here, is that embedded in Māori law is the concept of responsible water use, which sits above the right to use what we possess (Durie 2018). As detailed in the Whanganui River report, the relationship between Whanganui *iwi* and the river illustrates an ethic of care for the environment, which is more than a relationship based on ownership, possession and individual property rights. Part of the ethic of respect, and crucial to understanding Māori culture, is the significance of relationships. The relationship Māori sought in the encounter with the British and the signing of the treaty was one of trade, continual reciprocity and something closer to a partnership. For Europeans, the relationship was to seek out ownership and political control, and waterways and rivers were not exempt from these intentions.

After the appropriation and exploitation of rivers, lakes and streams in the 19th century, the New Zealand government looked for a clear and consistent policy for the management of natural resources (Durie 2018). This policy came in the form of the RMA 1991, which was considered revolutionary in legislative design for its integrated approach to planning for sustainable resource management (Jacobson et al. 2016). The legislation was designed to bring together multiple, often single-issue and even contradictory statutes under a single rationalized and integrated system (Joseph et al. 2019). The purpose of the RMA is to promote sustainable management of natural and physical resources through a hierarchy of plans and decision-making. The RMA states that all persons shall recognize and provide for amongst other things the relationship of Māori and their culture and traditions with their ancestral lands, water, sites, *waahi tapu* (sacred places), and other *taonga* (prized possession), as well as the protection of customary rights. All persons exercising functions under the RMA must also consider the principles of the Treaty of Waitangi.

Despite these stated intentions and safeguards, the interpretation and application of the RMA has been problematic for Māori for many reasons. One of which is the constraints and inability to exercise *rangatiratanga* because the RMA does not define Māori property rights regarding water (Memon and Kirk 2012). Māori concepts are often lost in translation, therefore integrating the principles of the treaty and *tikanga* (customs, practices or rules) into legislation is dependent on decision-makers, such as commissioners, regional and local councils, and the environment and high courts (Joseph et al. 2019). Often these decision-makers have little to no expertise or understanding of or connection with *mātauranga* (Māori knowledge) and *tikanga*. Whanganui *iwi* leader Gerrard Albert states that the RMA has developed a system to disassemble and invalidate our relationship with the lands and waters. An example of this is the emergence of cultural impact assessments, which are intended to provide a voice for Māori through documenting cultural values, interests and associations with a particular area or resource. Many assessments, however, are validated by outsiders with little experience in the local context and are seen to be a tick-box exercise, rather than an actual assessment through co-design and joint decision-making with Māori from the outset.

From the Whanganui River claims to Te Awa Tupua Act

Following the original filing of the Waitangi Tribunal 167 claim in 1990, the Whanganui River report was released in 1998. Negotiations for a just and durable solution for the Whanganui River commenced in 2002. From that point, the process of understanding the issues and crafting solutions based on the Whanganui River report took another 14 years and is reflected in four key documents. The first is the terms of negotiation signed in 2003 between the Crown and Whanganui *iwi*, which set the intentions and framework for negotiating a settlement for claims to the Whanganui River. The second document, the record of understanding, was signed in 2011 to outline the relationship between Whanganui *iwi* and the river. Two fundamental principles arose out of defining that relationship – the first is that Whanganui River is an integrated indivisible whole both physically and metaphysically, and second that there is an intrinsic link between the health of the river and the health of Whanganui *iwi*.

The third document, *Tūtohu Whakatupua*, was signed in 2012 between Whanganui *iwi* and the Crown and represents the parties' commitment to progressing the agreed arrangements for the settlement of the historical Treaty of Waitangi claims for the Whanganui River. The fourth and notably most important document, *Ruruku Whakatupua*, was signed in 2014. The signing of *Ruruku Whakatupua* between representatives of the Crown and Whanganui *iwi* at *Rānana Marae* on August 5, 2014 was a landmark step in the settlement of the historical Treaty of Waitangi claims for the Whanganui River. *Ruruku Whakatupua* is the Deed of Settlement for the Whanganui River and is made up of two documents; *Te Mana o Te Awa*, focusing on the health and wellbeing of the river and its status as Te Awa Tupua; and *Te Mana o Te Iwi*, which recognizes and provides for the relationship between Whanganui *iwi* and the river. At the core of the settlement is *Te Pā Auroa nā Te Awa Tupua*, which is the framework for Te Awa Tupua and includes the following components:

- The legal recognition of Te Awa Tupua as an indivisible and living whole comprising the Whanganui River from the mountains to the sea, incorporating its tributaries and all its physical and metaphysical elements.
- The legal recognition of *Tupua Te Kawa*, the intrinsic values of Te Awa Tupua.
- The appointment of *Te Pou Tupua*, to act as the human face of Te Awa Tupua.

- The development of *Te Heke Ngahuru ki Te Awa Tupua*, a whole of river strategy to guide the future health and wellbeing of Te Awa Tupua.
- The establishment of a collaborative group, *Te Kōpuka nā Te Awa Tupua*, comprising representatives of local and central government, Whanganui *iwi*, other *iwi* with interests in Te Awa Tupua and other key stakeholders, who will develop the *Te Heke Ngahuru* strategy.
- The vesting of the Crown-owned parts of the bed of the Whanganui River in Te Awa Tupua.
- A fund of $30 million dollars, *Te Korotete o Te Awa Tupua*, to support the health and wellbeing of Te Awa Tupua.

In 2017, Te Awa Tupua Act was passed and was the culmination of negotiations between Whanganui *iwi* and the Crown stretching back over 150 years. The Act puts the arrangements set out in *Ruruku Whakatupua* into legislation and includes legal personhood for the river. At the heart of the settlement legislation is better management systems for Te Awa Tupua, including unified policies, improved relational connections between nature and humans, and acknowledgement of the deep-seated relationship between Whanganui *iwi* and their river (Talbot-Jones and Bennett 2022). It is clear to see the long and arduous journey that Whanganui *iwi* have endured simply to exercise rights and responsibilities in relation to the Whanganui River in accordance with their *tikanga*. It is also clear to see that the components of Te Awa Tupua Act did not appear at the passing of the Act but rather have been with Whanganui descendants for centuries.

Te Awa Tupua Act and Te Tiriti o Waitangi

Although Te Awa Tupua Act came through a Treaty of Waitangi settlement process, the Act is framed around the river itself. As tribal leader Gerrard Albert iterates, treaty or no treaty, we would hold the same views about our place in the community and our lands and rivers. The courts, political parties, the government and their agencies, have treated the treaty as if only they can define what it means and how it should be applied. In fact, it should be defined by the community to whom it belongs – the people of Aotearoa New Zealand and in our context with our river, by the people of the river. Neither the Treaty of Waitangi nor *Te Tiriti o Waitangi* should be something that lives in the past. The treaty should be lived through the people at place and in real time; it is they who should decide what a fair, equitable and just society looks like. *Hapū, iwi* and Māori will always expect to have an equitable voice and decision-making authority and will seek the best outcome for the whole community, Māori and non-Māori (*Pākehā*).

The New Zealand government, through the Crown, will always change. How then, can you plan for and manage a river around a system that is constantly changing in terms of members of parliament and their respective political parties, policies and laws? The metaphor of a *pā auroa*, a broad eel weir, which can withstand the autumn, winter and spring floods, can be applied to the framework for Te Awa Tupua, in that it can withstand the constant changes being directed by central government. As a result, Te Awa Tupua creates permanency for river communities and ensures we are prepared, so that every three, six or nine years when a government wants to change the law, we can absorb that into a collaborative, cooperative, equitable community space that includes everybody and puts the river first. That community also involves *hapū* and *iwi*, who have a distinctive voice, and democracy exists to provide for our voices being heard. This is the contract that was signed in 1840 – we were not just meant to fold into or assimilate to the Crown's way of thinking or doing things.

Te Awa Tupua is an even stronger model than co-governance because it invites and requires everyone who has an interest in the river and who is impacted by it to work together. Where co-governance involves shared decision-making and arrangements between Māori and the Crown, Te Awa Tupua is better understood as a community-led model or community democracy. Te Awa Tupua Act commits everyone to the river in its entirety, without the need for the government to determine how things should be. As tribal leader Gerrard Albert continues, there is no exit clause; it is axiomatic in that no one can mandate or confirm us anymore under the RMA or any other legislation. Whanganui *iwi* stood on the banks and watched their eel weirs be destroyed, watched hydroelectric developments harm the river and watched farming practices pollute the water while having little power to do anything about it. In contrast, Te Awa Tupua Act is a more inclusive framework than what any other pre-determined treaty model could provide because it includes the community as part of the river's picture.

Te Awa Tupua Act and reinforcing a Whanganui perspective of our river

Given that Te Awa Tupua was the longest litigation case in the country, Whanganui *iwi* were constantly preparing and engaging themselves in processes and systems imposed by the state, such as court proceedings and cultural impact assessments. In doing so, Whanganui leaders believed they were continuously losing a sense of who they are at the expense of explaining and justifying themselves to others. The strategy for the settlement negotiations with the Crown sought to avoid this by simply leading with what Whanganui *iwi* knew, in particular our value system. During the design of the Te Awa Tupua framework, it was necessary to go back to our own stories, our own knowledges, our own narratives. By design, the legal agreements sought in the development of Te Awa Tupua Act not only mirror cultural precepts but reinforce those precepts (Takacs 2021). As lead Whanganui negotiator Gerrard Albert pointed out, Whanganui *iwi* set out to devise a different legal structure that accurately reflected and framed the worldview they already held.

The first part of this strategy was the need to move away from direct translation of Māori terms and concepts. *Tikanga* – a term that is used in much Aotearoa New Zealand legislation – can be taken to mean many things, such as customs, practices, rules or doing the correct thing. In Te Awa Tupua Act, however, the term *tikanga* is not used much. Instead, *kawa* is given the central role. This provides for a deeper integration of Māori law and values that encompasses *tikanga* as well as *mātauranga* (Māori knowledge), *kōrero tuku iho* (narratives handed down) and *whakapapa kōrero* (ancestral knowledge). While we can attempt to translate the term *kawa*, there is no English equivalent to it; therefore, it becomes difficult to define without the engagement of Whanganui *iwi*. The application of *kawa* is a more layered and sophisticated approach in that it focuses beyond reasonable practice towards asking questions of motive and intentions.

While Whanganui *iwi* negotiators wanted to reinforce a Whanganui *iwi* perspective of the river, others, especially non-Māori, saw this as creating a dichotomy between a Māori worldview and a Crown or Western worldview. Whanganui *iwi*, however, do not see anything wrong with the way we view, use and manage the river. Moreover, we believe the way we see the river is not only beneficial for the river itself but for the wider community. Te Awa Tupua Act acknowledges the transformative power of Indigenous knowledge and practice, now vested in legal personhood for the Whanganui River as a manifestation of an Indigenous paradigm of nature. Because it is a new and relatively unusual approach, most people see the novelty of applying legal personhood to a river only when it is explained in more depth, including the centrality of *Tupua te Kawa*.

As an example of the different worldviews, for more than 100 years the laws, regulations and actions of the Crown have broken the Whanganui River down into parts. The river is viewed and managed separately by the Crown in terms of its bed and banks, its subsoil, its waters, its flora and fauna and its fisheries; the river is subject to a raft of different statutory regimes including the RMA, the Local Government Act, the Fisheries Act, the Conservation Act, the Maritime Transport Act and the Marine and Coastal Area (*Takutai Moana*) Act; the river comes under the jurisdiction of one regional council, three local authorities and a number of Crown agencies; and issues of navigability, sale and compulsory acquisition mean the bed of the river and its tributaries are in parts privately owned and in other parts in Crown or public ownership (Ngā Tāngata Tiaki o Whanganui 2020b). Under Te Awa Tupua Act, the framework seeks to take a catchment-wide approach to ensure that all the waterways which together join to form the Whanganui River are viewed and managed not in isolation but with reference to the whole river as an interconnected ecosystem. This makes sense not only in cultural terms but also in environmental terms.

Te Awa Tupua Act and legal personhood

There is a lot of literature on Te Awa Tupua that focuses on the innovation of legal personhood, including varying interpretations of its purpose, the issues it addresses and how it is understood in the context of the Whanganui River. In our article (Cribb, Macpherson and Borchgrevink 2024), we also explain that in the literature the Te Awa Tupua Act is interpreted from a range of different ideological positions within the rights of nature orthodoxy, especially as evidence of an Indigenous origin to the rights of nature movement. We argue that some of these positions pay only superficial attention to Indigenous law or place-based activism and that such assessments must be made on the basis of thorough understanding of local nuances. We think it is useful here to provide some background on the origins of legal personhood and Te Awa Tupua. We refer to the book *He Kupu Taurangi: Treaty settlements and the future of Aotearoa New Zealand* written by Chris Finlayson and James Christmas. Finlayson was the Minister for Treaty Settlements during the Whanganui River negotiations. Finlayson, a lawyer by trade, begins by stating that given previous debates on water ownership, transfer of ownership of the Whanganui River was a step too far for several other members of parliament. Many of those colleagues also felt that it would also be a step too far for the wider public. Legal personhood came about by John Wood, who in 2009 was appointed as the Crown's lead negotiator for the Whanganui River case. Wood was interested in legal fictions and situations where the law recognizes personality in a non-human, living or non-living object.

Wood's first negotiation was with a tribe called *Ngai Tuhoe*, which involved taking a national park, an area known as Te Urewera, out of Crown ownership and vesting the land in itself (Finlayson 2019). While some might assume this led to Whanganui River being afforded legal personhood, it was following Whanganui *iwi* suggesting a unique title for the river that legal personhood was proposed to *Ngai Tuhoe* as a solution for Te Urewera National Park. Finlayson describes precedents of legal personhood in his book (Finlayson and Christmas 2021), including the much-quoted article by Christopher Stone – which considered whether trees should have standing (Stone 1972), legal personhood for Indian temples, and the Waikato River being represented by a guardian – and Tony Angelo's article about *taonga* or prized possessions given a form of legal personality (Angelo 1996). Professor Angelo's article is specific to Aotearoa New Zealand, especially his observation of legal personality in the context of the treaty as a bridge between two largely incompatible worldviews.

The literature on Te Awa Tupua states the varying issues facing the Whanganui River that legal personhood has become the solution to and where it falls short. Legal personhood was particularly proposed in the first place as a solution to ownership of freshwater being off the table (Takacs 2021). However, as Sanders (2017) argues, why should ownership matter when the thing that is owned is known to have been acquired unjustly? This argument refers to the issue discussed earlier of ownership and property rights, exacerbated by the fact that many New Zealanders have little knowledge of the way lands and rivers were illegally acquired by the Crown. Through legal personhood, it was more acceptable for a river to own itself, if it could not be owned by Māori or the Crown. This could be seen as a political sacrifice by Whanganui *iwi* given the Te Awa Tupua Act does not have any impact on existing property rights or rights to or interests in water. As tribal leader Gerrard Albert reminds us, ownership does not consider the totality of the relationship between people and nature because, essentially, you cannot own something you belong to.

Another perspective in the literature sees legal personhood as a reset of the relationship between people and nature, as well as regulating the relationships between Māori and the Crown (Sanders 2017). This reset includes the potential for how non-Māori can contribute to the health and wellbeing of the river through the Te Awa Tupua Act. For example, if the Act promotes a more collaborative and cooperative approach to understanding and using the Whanganui River, this requires the community, key stakeholders and government agencies to work differently than they have before. The relationship reset therefore requires an unpacking of previous thought patterns and processes, especially those guided by the RMA. In fact, it was necessary to involve relevant local authorities in the early stages of designing the Act; input at a technical level ensured they would not react or disagree to arrangements once the Act became legislation (Ngā Tāngata Tiaki o Whanganui 2016). The relationship reset also mandates collaboration between previously antagonistic parties (Finlayson and Christmas 2021). For example, energy companies will have to talk to environmental and recreational groups because there is now every incentive for everyone to work together without time consuming and expensive litigation. Furthermore, Finlayson goes beyond a relationship reset to suggest the compulsory need for people to change, which can be advanced by the application of legal personhood. It is uncertain whether people would voluntarily change harmful practices or attempt to see nature and rivers in a different way without the encouragement of the law. This encouragement needs to be managed delicately, however, so that the Te Awa Tupua Act does not have the opposite effect and become a legislation of extreme enforcement.

Legal personhood is also recognized as a solution to matters of policy fragmentation, adversarial relationships and failings of current legislation, in that it reflects the values shared by Whanganui *iwi* (Talbot-Jones and Bennett 2022). It is argued that through the Te Awa Tupua Act, Whanganui *iwi* can now have greater participation in affairs relating to the river, in terms of being able to express the deep-seated beliefs they hold about their river. Through legal personhood, the settlement of the Whanganui River claims concluded 150 years plus of grievance by enshrining in legislation the tribe's perspective and the importance of the river to them (Finlayson and Christmas 2021). Enshrining a tribal perspective in law means Indigenous, customary law is now being upheld by Westminster law through recognizing the sovereignty and legacy of the river itself. Whanganui *iwi* leaders, however, have framed legal personhood as the closest approximation to the worldview held by Whanganui *iwi*. As Hawira (2019) reminds us, legal personhood is the vehicle to carry the new status, nothing more, meaning that Whanganui *iwi* rooted their negotiation strategy with the Crown in the traditional Māori relationship with the river (Takacs 2021). Regardless of how legal personhood is perceived, Whanganui *iwi* remind us that it is still a legal fiction that does not belong to Whanganui *iwi*, and at the core of Te Awa Tupua Act is *Tupua te Kawa*, the value system for the Whanganui River.

Since the passing of Te Awa Tupua Act

In 2024, Whanganui *iwi* celebrated 10 years since the signing of *Ruruku Whakatupua*, the Deed of Settlement at *He Pukenga Wai*, the Whanganui River symposium. On day two of the symposium, the Riverine Rights project presented their findings as one of the first collaborations between Ngā Tāngata Tiaki (the Whanganui River post-settlement governance entity) and an international research project. The focus of day two was sharing collaborative endeavours within the community to nurture and advance the health and wellbeing of Te Awa Tupua. Key findings from our research are comprehensively explained in our article "Beyond legal personhood for the Whanganui River: Collaboration and pluralism in implementing the Te Awa Tupua Act" (Cribb, Macpherson and Borchgrevink 2024). By analysing two cases, we offer new perspectives of how legal and governance frameworks for the Whanganui River are being implemented. The first case is *Te Pūwaha*, a collaborative community-led project, partnering with the Provincial Growth Fund, to restore and revitalize the Whanganui Port precinct including infrastructure. The second case study, *Te Kōpuka nā Te Awa Tupua*, is the 17-member collaborative group charged with developing *Te Heke Ngahuru*, the Whanganui River strategy. Through an examination of place-based examples, we illustrate a shift away from the legal personality to the centrality of *kawa*, the establishment of a set of new institutions and practices to uphold *kawa*, and the devolution of authority to *hapū* and communities.

When we first wanted to look at case studies of "implementation" of the Te Awa Tupua Act, we first had to decide what we meant by implementation. What first appeared as a rather technical, even transactional term we wanted to understand how the Act was being understood, practised, applied and potentially embedded by the people at place, namely Whanganui *iwi* and communities. We decided on cases where there had been enough experiences to reflect on and public material to refer to and cite. *Te Pūwaha* is considered the first major infrastructure project on the Whanganui River under the Te Awa Tupua Act, and there were certainly many people to speak to for this case study. *Te Kōpuka nā Te Awa Tupua* was suitable for other reasons in that it is not a project as such, but rather a core component of the Te Awa Tupua framework and a suitable case study that involves many different individuals.

At the heart of the two cases of *Te Pūwaha* and *Te Kōpuka nā Te Awa Tupua* are the priority, centrality and elevation of *kawa*, reaffirming Indigenous jurisdiction through *Tupua te Kawa*. Part of centralizing *kawa* represents a paradigm shift and significant departure from the old RMA way of doing things. Examples include the move from schematic consultation – that is, seeing *iwi* or *hapū* as a bus stop at the end of a process to tick a box, to collaboration, by bringing everyone into the conversation to do with the river. The paradigm shift also includes moving from a transactional to a relational approach. A transactional approach, such as the consenting and cultural impact assessment processes, is often short-term focused, motivated by immediate returns lacking genuine connection and emotional investment. A relational approach, on the other hand, is based on building trust and mutual respect through continuous dialogue over time – picking up the phone and having a conversation when the need arises, arranging meetings, preferably face to face, and building reciprocal relationships. While this relational process of engaging everybody involved may take more time initially, it leads to a more democratic process and outcomes that are community-owned and broadly accepted. For *Te Pūwaha, Tupua te Kawa* guides all decision-making and operations in the project, and as part of the objectives for *Te Heke Ngahuru*, the Whanganui River strategy, all decisions of *Te Kōpuka* are based on *Tupua te Kawa*.

As part of centralizing *kawa*, there needs to be systems and structures in place to support and uphold *kawa*, and our research indicated several new institutions and practices that have been established to uphold *Tupua te Kawa*. For *Te Pūwaha*, this is seen in the development of *He Ara Tuku Rau*, an assessment framework to support decision-making and an operationalization pathway for the implementation of the Te Awa Tupua Act including *Tupua te Kawa. He Ara Tuku Rau* promotes the return to an original state of abundance for the river, which is quite different from trying to avoid, remedy and mitigate adverse environmental effects, which are the objectives of the RMA. As part of the process to socialize and ensure everyone understands *Tupua te Kawa*, a series of workshop inductions for employees, contractors and their staff were conducted. For *Te Kōpuka*, considerable time –three and a half years – was spent on relationship building for the group to get to know each other. This included the development of safe spaces to air their views and creating a common understanding of the issues faced by communities as well as an understanding of *Tupua te Kawa*.

As part of the creation of new institutions, new entities are created, some of which are provided for by the legislation, such as *Te Kōpuka nā Te Awa Tupua*. Others are established for specific purposes and contexts. *Te Mata Pūau* is a *hapū* (sub-tribe) -led collective overseeing the application of the Te Awa Tupua Act with key partners at both the governance and operational levels of *Te Pūwaha* project. Of the 17 members of *Te Kōpuka nā Te Awa Tupua*, the strategy group, six appointments represent the *iwi* with interests in Te Awa Tupua. *Te Ripo* was established to facilitate an *iwi* presence, required for the appointment of the *iwi* membership of *Te Kōpuka* and is currently represented by 11 *iwi*, of which Whanganui *iwi* is one. While these examples illustrate *iwi* and *hapū* involvement, involvement by the non-Māori community has been just as crucial, especially given the large presence of non-Māori that live in the Whanganui city and along the river. As well as *Te Pūwaha* bringing together a broad constellation of actors – tribal organizations, a private company and public entities – the project governance board also includes local community representation. *Te Pūwaha* have also attempted to involve the wider community through a series of open meetings and events to explain proposed activities. While not all residents have agreed with developments, there has been greater opportunity for engagement than there was under the RMA. For *Te Kōpuka*, it is important to note that local authorities only represent four of the 17 positions, with the remaining membership sitting within the communities.

Through both cases, we can see the centrality and relevance of not only how *Tupua te Kawa* is understood but how it is applied, shifting the framing from the legal fiction of the person towards the broader enabling jurisdiction for *kawa*. There has also been a shift away from limiting the damages of using nature as a resource under the RMA towards *kawa*-based notions of relationality and abundance. To uphold and support the *kawa*, a set of new institutions and practices had to be established, ones which devolve authority to *hapū* and communities. In our article (Cribb, Macpherson and Borchgrevink 2024), we conclude that the Te Awa Tupua Act is a recognition of relational legal pluralism. This recognition is not in the sense of static declarations of rights or arriving at a state of finality but recognition of the ongoing relationships between Indigenous law and state law, and people and the river through *whakapapa* or genealogical constructs and relationality.

What we have also learnt across the two cases is that this new way of working has not come without its challenges. These include understanding if and how Te Awa Tupua Act works within or alongside other pieces of legislation, as stated in the above paragraph on the relationship between Indigenous and state law. In our interviews, there were variances in the levels of interpretation of what this means and whether and how the Act sits alongside other pieces of legislation, despite specific clauses in the Act stipulating how this is done. Other people we spoke to raised issues of the meaning and application of terms in the Act such as "recognize" and "provide for", which some felt did not go far enough in terms of exercising authority and protection for the river. Further conceptual and practical challenges were raised, such as community buy-in through ongoing socialization and understanding, and the time and extended costs to build meaningful relationships. The benefits in the long run from better and more inclusive planning would outweigh up-front costs by far and is part of the reason why commitment to long-term benefits is necessary. Across both cases we have also learnt that initiatives for the Whanganui River are context specific, and what you might do at *Te Pūwaha* is not necessarily what you would do elsewhere along the river. Just as the environmental degradation of the river occurred over decades, change will be incremental. Nonetheless, both cases send a signal of what is expected when working towards the betterment and health and wellbeing of the river and how this can be achieved through the Te Awa Tupua Act. Finally, we are yet to see the full influence of the Te Awa Tupua Act on future resource consenting processes, especially regarding whether the concession for the diversion of water through the Tongariro Power Scheme will be renewed in its present form. It is premature, therefore, to say that the journey for Whanganui *iwi* to regain sovereignty over the river is fully over.

Conclusion

This chapter provides a layered understanding to the story of Te Awa Tupua by situating it in its historical, political, social and cultural context. We have described the geography of the river and the values and practices held by descendants, both historically and today. Whilst we hold the same values and practices concerning our river as those of our ancestors, physically the river is in a much more dire state than it was previously. As a result of Crown breaches against the treaty, Whanganui *iwi* have used the treaty settlement process as a legislative attempt to gain recognition of their sovereignty over the river and correct the wrongs inflicted by colonization. After the culmination of over 100 years of legal proceedings, legal personhood became the mechanism to address this issue, inadvertently contributing to the varying conceptual and ideological positions within the rights of nature orthodoxy. These positions include legal personhood as a solution to water ownership, thus resetting the relationship between people and nature. There are also positions of understanding legal personhood as a bridge to different worldviews, therefore recognizing Indigenous knowledge. Some argue that legal personhood allows for greater participation by *hapū* and communities, while also making it compulsory by law for people to change and manage relations with the river differently.

Others, however, say that Te Awa Tupua does not go as far as to address existing property and water rights, resulting in a political sacrifice by Whanganui *iwi*. Furthermore, Whanganui *iwi* leaders remind us that the ownership argument is problematic because ownership does not consider the totality of the relationship between Whanganui *iwi* and the river. And although legal personhood may provide the closest approximation to a Whanganui *iwi* worldview, it is still a legal fiction that exists to carry the new status of the river. At the heart of the Act are the *kawa* or Indigenous laws, representing the need for a paradigm shift and for the institutions to support and ensure the health and wellbeing of the river. For these reasons, we would argue that Te Awa Tupua is first and foremost an Indigenous law and rights framework because it entrenches an Indigenous viewpoint of the river which sees people as part of nature. However, this is not in contrast to or at the expense of or above our responsibilities to the rights of nature. As we have argued in our article (Cribb, Macpherson and Borchgrevink 2024), Te Awa Tupua is an Indigenous sharing model by virtue of the fact that through relationality, abundance, working together, putting the river first and bringing a whole community in conversation with the river, the Te Awa Tupua model benefits everyone.

All these positions contribute to an increasing level of nuance, broadening our conceptual and theoretical understanding of Te Awa Tupua, especially in the context of riverine rights and rights of nature. Together with the motives and processes that led to the passing of the Act, we have a better understanding of why legal personhood, why now and, to a lesser extent, why rivers, noting the previous granting of personhood in Aotearoa New Zealand to a national park and, soon to be, a mountain. What remains is the extent to which the Te Awa Tupua Act can address environmental degradation of the river and the practicalities and potential of the Act to help inform the health and wellbeing of other rivers. What Whanganui *iwi* leader Gerrard Albert reminds us though is that given the novelty of what has been achieved with Te Awa Tupua it is easy to get caught up in the semantics or what we noted earlier as the symbolism of the Act. That is, trying to quantify what the Te Awa Tupua Act intends to do, will do or not do. This is further challenged by a need to avoid generalizations and binaries in our conceptualizations, but rather appreciate each case in its own current context. Given the learnings shared of our place-based experiences, it will be imperative to focus our efforts on long-term continuation by directing our intentions and responsibilities towards restoring the health and wellbeing of nature, of rivers – in this case, the Whanganui River.

References

Angelo, A. 1996. "Personality and legal culture." *Victoria University of Wellington Law Review* 26 (2): 395–412. doi:10.26686/vuwlr.v26i2.6174.

Anderson, Atholl, Judith Binney and Aroha Harris. 2014. *Tangata Whenua: An illustrated history.* Bridget Williams Books.

Andrew, Jessica. "Administrative review of the Treaty of Waitangi Settlement Process. 2008." *Victoria University of Wellington Law Review* 39: 225–264.

Bargh, Maria, ed. 2020. *Diverse indigenous environmental identities.* Routledge.

Cribb, Miriama, Elizabeth Macpherson and Axel Borchgrevink. 2024. "Beyond legal personhood for the Whanganui River: Collaboration and pluralism in implementing the Te Awa Tupua Act." *The International Journal of Human Rights*: 1–24, doi:10.1080/13642987.2024.2314532.

Crocker, Therese. 2014. "History and the Treaty of Waitangi Settlement Process." *Journal of New Zealand Studies* 18: 106–117.

Durie, Edward Taihākurei, ed. 2018. *Indigenous law and responsible water governance.* Routledge.

Finlayson, Chris, ed. 2019. *A river is born: New Zealand confers legal personhood on the Whanganui River to protect it and its native people.* CRC Press.

Finlayson, Chris, and James Christmas. 2021. *He Kupu Taurangi: Treaty settlements and the future of Aotearoa New Zealand.* Huia Publishers.

Harmsworth, Garth, and Shaun Awatere, eds. 2013. *Indigenous Māori knowledge and perspectives of ecosystems.* Manaaki Whenua Press.

Hawira, Turama. 2019. "Te Pou tupua, voice of the Whanganui River." Presentation at the Second Annual Assembly of First Nations National Water Symposium, February 26–28, Niagara Falls, Ontario. https://81e28e4e-2cec-40db-a85d-6c7aa98a928e.filesusr.com/ugd/89afed_39c2d2beb64d4942ac286a5eca88ebc3.pdf.

Hayward, Janine, and Nicola Wheen. 2004. *The Waitangi tribunal: Te Roopu Whakamana I Te Tiriti O Waitangi.* Bridget Williams Books.

Jacobson, Chris, Hirini Matunga, Helen Ross and R.W. (Bill) Carter. 2016. "Mainstreaming Indigenous perspectives: 25 years of New Zealand's RMA." *Australasian Journal of Environmental Management* 23 (4): 331–337. doi:10.1080/14486563.2016.1259201.

Joseph, Robert, Mylene Rakena, Mary Te Kuini Jones, Rogena Sterling and Celeste Rakena. 2019. "The Treaty, Tikanga Māori, ecosystem-based management, Mainstream law and power sharing for environmental integrity in Aotearoa New Zealand – Possible ways forward." https://www.sustainableseaschallenge.co.nz/assets/dms/Reports/The-Treaty-tikanga-Maori-ecosystem-based-management-mainstream-law-and-power-sharing-for-environmental-integrity-in-Aotearoa-New-Zealand-possible-ways-forward/MAIN20TuhonohonoSSeas20Final20Report20Nov202019.pdf.

Kovach, Margaret. 2009. *Indigenous methodologies: Characteristics, conversations, and contexts.* University of Toronto Press.

Memon, Pyar Ali, and Nicholas Kirk. 2012. "Role of Indigenous Māori People in collaborative water governance in Aotearoa/New Zealand." *Journal of Environmental Planning and Management* 55 (7): 941–959. doi:10.1080/09640568.2011.634577.

Mutu, Margaret, ed. 2004. *The Humpty Dumpty Principle at work: The role of mistranslation in the British settlement of Aotearoa.* Routledge.

Mutu, Margaret. 2018. "Behind the smoke and mirrors of the Treaty of Waitangi Claims Settlement Process in New Zealand: No prospect for justice and reconciliation for Māori without constitutional transformation." *Journal of Global Ethics* 14 (2): 208–221.

Ngā Tāngata Tiaki o Whanganui. 2016. "Submission to Māori Affairs Select Committee: Te Awa Tupua (Whanganui River Claims Settlement) Bill." https://www.ngatangatatiaki.co.nz/assets/Uploads/Important-Documents/NTT-Submission-Te-Awa-Tupua-Whanganui-River-Claims-Settlement-Bill.pdf.

Ngā Tāngata Tiaki o Whanganui. 2020a. "Historical Journey." https://www.ngatangatatiaki.co.nz/our-story/historical-journey/.

Ngā Tāngata Tiaki o Whanganui. 2020b. "Te Pā Auroa Nā Te Awa Tupua." https://www.ngatangatatiaki.co.nz/our-story/ruruku-whakatupua/te-pa-auroa-na-te-awa-tupua/.

Orange, Claudia. 2015. *An illustrated history of the Treaty of Waitangi.* Bridget Williams Books.

Pākaitore Historic Reserve Board. 2020. *Pākaitore: A history.* H & A Print.

Sanders, Katherine. 2017. "'Beyond human ownership'? Property, power and legal personality for nature in Aotearoa New Zealand." *Journal of Environmental Law* 30 (2): 207–234. doi:10.1093/jel/eqx029.

Stone, Christopher. 1972. "Should trees have standing: Toward legal rights for natural objects." *California Law Review*: 450–493.

Takacs, David. 2021. "We are the river." *University of Illinois Law Review* 2: 545–605. https://repository.uchastings.edu/cgi/viewcontent.cgi?article=2830&context=faculty_scholarship.

Takiari, Kiritahi. 2002. *Women of the river: Part one.* ICT Whanganui Charitable Trust.

Talbot-Jones, Julia, and Jeff Bennett. 2022. "Implementing bottom-up governance through granting legal rights to rivers: A case study of the Whanganui River, Aotearoa New Zealand." *Australasian Journal of Environmental Management* 29 (1): 64–80.

Tanasescu, Mihnea. 2022. *Understanding the rights of nature: A critical introduction.* Transcript Verlag.

Tinirau, Rawiri, Cherryl Smith, Meri Haami and Helena Rattray-Te Mana. 2020. *Te Oranga O Te Awa Tupua: A report prepared for Ngā Tangata Tiaki O Whanganui Trust*. Te Atawhai o Te Ao Charitable Trust.

Turia, Tariana. 2001. "The Whanganui River Claim – The cultural and spiritual significance to the Iwi of the Awa." https://www.scoop.co.nz/stories/PA0110/S00149.htm.

Waitangi Tribunal. 1999. *The Whanganui River Report*. GP Publications.

Waitangi Tribunal. 2014. *He Whakaputanga Me Te Tiriti – The declaration and the treaty: The report on Stage 1 of the Te Paparahi O Te Raki inquiry.*

Waitangi Tribunal. 2020. "Claims Process." https://waitangitribunal.govt.nz/claims-process/.

Wheen, Nicola R., and Janine Hayward, eds. 2012. *Treaty of Waitangi Settlements.* Bridget Williams Books Limited.

Zartner, Dana. 2021. "Watching Whanganui and the lessons of Lake Erie: Effective realization of rights of nature." *Vermont Journal of Environmental Law* 22 (1): 1–39. https://heinonline.org/HOL/LandingPage?handle=hein.journals/vermenl22&div=4&id=&page=.

3 The biocultural rights of the Atrato River, Colombia

Catalina Vallejo Piedrahíta, John Andrew McNeish, Jenny Moreno Socha, Arlinton Cuesta Mosquera and Daniel Alfonso León

Introduction

The Atrato River, a vital waterway in Colombia, flows through the ecologically rich Chocó region on the country's Pacific coast. The river is central to the livelihoods, cultural practices and social structures of the Afro-Colombian and Indigenous communities who reside along its banks. For these communities, the river is more than a physical entity – it is a profound symbol of their identity, history and continued survival. However, the region's natural wealth has attracted activities such as mining and deforestation, which have introduced significant pollution and environmental degradation. The contamination of the river's waters, particularly from heavy metals linked to illegal mining, has led to serious health and wellbeing challenges for the communities that rely on it for drinking water, fishing and agriculture.

Spanning nearly 750 kilometres, the Atrato River is Colombia's third-largest river and borders the Chocó Biogeographic Region, one of the planet's recognized biodiversity hotspots (Aguirre Ceballos and Rangel 2008). This region is located between the Pacific Ocean and the Western Andes, spanning territories of what are now the departments of Chocó, Antioquia, Valle del Cauca, Cauca and Nariño. The Chocó Department is home to a population of approximately 500,000 individuals. Of this population, 87% are of Afro-Colombian descent, 10% are Indigenous and 3% are mestizo (mixed European – typically Spanish – and Indigenous American ancestry) (Gobernación del Chocó 2020, Poveda et al. 2004).

In Colombia's legal framework, ethnic communities have achieved legal recognition of their collective rights through resistance and struggle. The Political Constitution recognizes Colombia as a pluri-ethnic and multicultural state (Colombian Constitution 1991, arts. 1 and 7, see Constituent Assembly of Colombia 1991). Ethnic communities are legally entitled to self-determination and self-governance, which grants them the authority to make decisions regarding their social, cultural and economic affairs. The decisions of the Inter-American Court of Human Rights have been significant in recognizing Indigenous and tribal peoples' rights, as seen in the Saramaka People v. Suriname (IACtHR 2007) and Case of the Kaliña and Lokono Peoples v.

DOI: 10.4324/9781003581208-3

Suriname (IACtHR 2015). Additionally, the International Court of Justice's advisory opinion on Western Sahara (ICJ 1975) and the International Labour Organization's Convention No. 169 on Indigenous and Tribal Peoples (ILO 1989) – which defines tribal peoples as communities with their own distinctive cultures and social institutions that existed prior to colonization – provide foundational legal frameworks for Indigenous rights protection. Complementing these are regional instruments such as the African Charter on Human and Peoples' Rights (OAU 1981), the American Declaration on the Rights of Indigenous Peoples (OAS 2016) and the United Nations Human Rights Committee's General Comment No. 12 on the right to self-determination (HRC 1984), which further reinforce the recognition and protection of collective rights of Indigenous and tribal communities.

The Chocó region is characterized by its *collective territories*. Collective territories are lands collectively owned and managed by Indigenous groups or Afro-descendant community councils, where they have the right to govern and protect their resources according to their traditions and cultural practices.[1] These communities, many of whom trace their ancestry to those who sought refuge during the era of slavery, share deep cultural and spiritual connections to the river, which plays a central role in their cultural and social practices (Pérez-Escobar et al. 2019). In the region, there are 683 Afro-Colombian communities organized into 64 Community Councils, as well as 125 Indigenous reserves of the Embera-Dóbida, Embera-Katío, Embera-Chamí, Wounan and Tule Indigenous ethnic groups (Gobernación del Chocó 2020, 36).

The communities of the Atrato River basin face persistent challenges, such as limited access to education and healthcare, social exclusion and the detrimental effects of resource extraction (García Reyes 2014, Palacios-Torres et al. 2018, Universidad de Córdoba and Universidad Tecnológica del Chocó 2022). In 2015, organizations representing these communities petitioned the Court to take action to resolve breaches to their fundamental human rights caused by mining and logging.

In 2016, the Colombian Constitutional Court responded with the landmark *Judgment T-622*, which recognized the violation of "biocultural rights" in the Atrato River basin. As defined in the ruling, biocultural rights emphasize the interconnectedness of biological and cultural diversity, ensuring local communities' ability to maintain and manage their territories based on traditional knowledge and practices. In using the framework of biocultural rights, the ruling highlighted that the protection of ecosystems is inseparable from the cultural and social rights of the communities that depend on them. This decision marked a shift in legal thought in the country, declaring the plaintiffs' human rights violated whilst also recognizing the Atrato River itself as a subject of rights, with the right to protection, preservation, maintenance and restoration. Grounded in Colombia's 1991 Constitution – which enshrines environmental protection and the recognition of ethnic and cultural diversity (Colombian Constitution 1991, art. 7) – the ruling legally recognized the deep

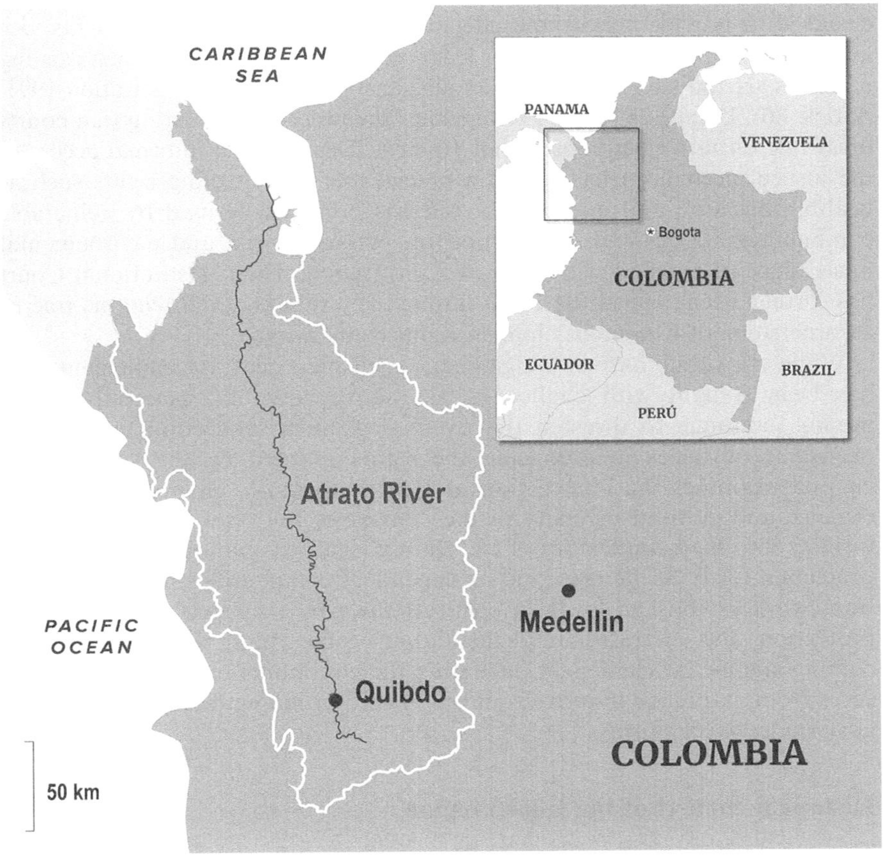

Figure 3.1. The Atrato River Basin. Map data ©2025 Google, INEGI. Elaborated by Daniel Alfonso León.

connection between the river and its dependent communities. To ensure enforcement, the Court established a commission of river guardians, composed of local community representatives and state institutions, tasked with overseeing the river's rights and ensuring participatory decision-making. Additionally, the Court mandated actions to address pollution, initiate ecological restoration and develop indicators to monitor the river's health, among other mandates (Judgment T-622, see Constitutional Court of Colombia 2016).

The Atrato case was brought to court through a *tutela* (constitutional writ of protection). In the Colombian legal system, a *tutela* is an informal legal mechanism that allows individuals to seek immediate protection of their constitutional rights when they are threatened or violated by public or private entities. Established under Article 86 of the 1991 Constitution, the *tutela* is designed to provide a swift and effective remedy for fundamental rights violations, particularly when no other legal recourse is available or when existing remedies are too slow to prevent serious harm (Colombian Constitution 1991, Article 86). Decree 2591 of 1991 regulates the procedure, ensuring that courts must rule on *tutela* petitions within 10 days. This fast and informal access to the justice mechanism has played a crucial role in protecting rights such as health, education and due process and has been widely used by vulnerable communities to demand state action on pressing social and environmental issues since its creation with the new Constitution. The Constitutional Court has further strengthened the *tutela* through key rulings, reinforcing its role as a cornerstone of Colombia's human rights framework.

While the ruling has set a significant legal precedent, its implementation has been fraught with challenges. Extractive activities, especially illegal mining, continue to threaten the river's ecosystem, and communities face persistent obstacles in exercising the rights granted by the Court. This chapter examines the intersection of legal innovation, environmental governance and cultural diversity in the context of the Atrato River, offering insights into the complexities of translating legal recognition into actionable protection for both the river and its dependent communities. While the legal framework established by the Constitutional Court offers a foundation for protection, the successful implementation of the ruling will require overcoming significant challenges, including the continuing environmental pressures from extractive industries and the need for strengthened collaborative governance mechanisms.

Historical context of the Chocó region

The specific history of the Atrato River basin is reflected in the experiences lived by Afro-descendants, Indigenous people and mestizos, whose circumstances and conditions over time reveal how their existence has been transformed. These ways of life reflect interpretations of the world and a relationship with the environment through a set of everyday practices, which are the result of their history, their position in the surrounding society and their form of social reproduction. Before the Spanish occupation, the Emberá and Cuna peoples had a long history of both alliances and conflicts as they moved through the basin. According to oral tradition, the Emberá worked with gold, lived around immense trees and began occupying the territories of the Cuna (Vargas 1990, 81).

Santa María la Antigua del Darién (1524), Acla (1541) and San Sebastián de Urabá (1552) were among the first Spanish settlements established to support the missions, the 'reduction of Indigenous people' (*reducciones de indios*),[2] and the extractive, commercial and mining economy. Expeditions, brigades and the establishment of mining districts were carried out from the provinces of Popayán and Antioquia towards the Atrato River lands. These relied on Indigenous labour and techniques for gold extraction, resources that helped the monarchy with its trade and wars in Europe and the Mediterranean (Rodríguez Hernández 2014, 67).

The Spanish incursions into the Atrato River basin led to a decline in the Indigenous population and a sudden change in their relationships: some groups disappeared, others were displaced and some resisted, altering their ways of life. After the second half of the 17th century, Chocó continued to be a focal point for mining exploitation, and a new strategy of occupation emerged. The Indigenous communities continued to be concentrated into settlements, while, due to a lack of labour, the Spanish began using African slaves and Creoles (people of European – usually Spanish – descent who were born in the Americas) to work the alluvial deposits of the river basins (Leal León 2016, 154).

During the 18th century, the intensification of slavery, along with the decline of the Indigenous population, resulted in a strong Afro-descendant presence. The Atrato and San Juan rivers were (and still are) centres of gold extraction, which largely relied on forced African labour. In 1782, it was estimated that there were 7,088 enslaved Afro-descendants, 340 whites (mostly absent or living outside the territory) and 4,000 free Black people. Slaves could achieve freedom by buying themselves free, or through voluntary manumission by slave owners as a way of cutting costs during periods of low mining profitability. According to William Sharp (1976), by the end of the 18th century and the beginning of the 19th century, free Black people outnumbered slaves.

The Afro-descendants of the Atrato, San Juan and Baudó rivers not only engaged in fishing and agriculture for the regional market but also continued gold extraction and the collection of rubber, *tagua* (vegetable ivory), wood and *níspero* (a type of tropical fruit). In the 750 km stretch from Quibdó to Cartagena – between the Pacific Ocean and the Andes – various groups gradually settled along the river, including Indigenous villages, slaves, 'freedmen' (former slaves who were freed) and 'maroons' (escaped slaves who formed independent communities) (Leal León 2016).

Over time, Black and Indigenous communities have developed cultural and productive systems that have been passed down through generations. These communities have woven a network of ongoing exchanges, both internal and external, including the exchange of goods between families (such as food, tools and gifts) and of services such as collaborating in activities like gold washing, clearing land or building homes. They also established trade with shops and commercial houses in Quibdó and Cartagena. *Champas*, boats similar to canoes, became the main means of transport, moving in and out of the Afro-descendant, mestizo and Indigenous towns and villages via the rivers and thus connecting them (García Reyes 2014). After Colombia's independence in the early 19th century, the landowning order in the province of Chocó began to weaken. However, racial stereotypes persisted, and the region's perceived underdevelopment was widely attributed to its ethnic composition and isolation from the rest of the country (Almario García et al. 2015).

The 19th century also brought significant legal advancements, such as the 1821 Law on the Gradual Abolition of Slavery, which marked an initial step towards emancipation. In the Atrato River basin, Afro-descendant families began shaping new economic, social and cultural dynamics, taking advantage of new opportunities to build alternative ways of life.

> With the dissolution of slavery, the decline of 18th-century gold mining settlements, the redefinition of Indigenous towns, the process of Independence and the formation of the Republic, and the growth of the population, conditions emerged for an extensive settlement along the Atrato River and its main tributaries. This demographic and social transformation was led primarily by Afro-descendant communities, though it also involved the reorganization of Indigenous groups.
>
> (Almario García et al. 2015, 51, our translation)

Between the 20th and early 21st centuries, the settlement process along the Atrato River evolved in multiple directions. Foreign merchants and landowners established a new land-based power structure, shifting from mining to a system based on extractive commerce. Meanwhile, for free Afro-descendant families, gold mining remained central to their scattered settlement patterns and territorial expansion strategies. At the same time, agriculture gained importance, particularly the commercialization of rubber and *tagua* (palm tree nut). These mining and agricultural dynamics fuelled steam navigation along the Atrato, facilitated the construction of roads connecting Chocó to Antioquia, Quindío and Cauca, and contributed to the consolidation of Quibdó as the capital and economic centre of Chocó (Jiménez Meneses 2001).

The Atrato basin today: A contested landscape of rights and resources

The Atrato River's ecological diversity, cultural significance and strategic location between two oceans have made it a contested and exploited region. Since the 2000s, the neoliberal policies of the Colombian government, including expanded mining rights for foreign companies and infrastructure projects, have exacerbated pressure on the region's ecosystems and deepened the vulnerabilities of local Afro-Colombian and Indigenous communities (Centro Sociojurídico para la Defensa Territorial Siembra 2022). Extractive industries in the river basin linked to formal and informal global trade networks – such as mining, banana and oil palm plantations, cattle ranching, logging and coca cultivation – have caused severe ecological and social harm (Tubb 2020, Centro Sociojurídico para la Defensa Territorial Siembra 2022). Over the past three decades, territorial conflicts and the presence of illegal armed groups have further destabilized the area (García Reyes 2014).

Illegal mining is the primary driver of environmental damage in the Atrato basin. Using unregulated machinery and mercury to extract gold, these operations have contaminated water sources, devastated fish populations and endangered public health. Mercury bioaccumulation in fish has exposed Afro-Colombian and Indigenous communities to severe health risks, including neurotoxicity (Kerguelén et al. 2022, Palacios-Torres et al. 2018, Universidad de Córdoba and Universidad Tecnológica del Chocó 2022). The illegal mining is controlled in large degree by organized criminal groups such as *Clan del Golfo* and guerrilla organizations such as the National Liberation Army (ELN). The illegal mining scene in Chocó is marked by the involvement of these armed groups, who not only control and profit from illegal mining but also extort miners. This extortion is a significant burden on the local miners, who are often forced into these situations due to the lack of state regulation, the lack of state presence (police and military) and the dominance of armed actors in the region. Corruption within state institutions at the local and regional level facilitates the continuation of illegal mining by providing information to illegal miners about military operations against them. This situation exacerbates the challenges of curbing illegal mining in the region (Bonza 2024).

Beyond mining (principally for gold, but also for sand and emeralds), other extractive activities – including logging, monoculture farming and coca cultivation – have contributed to ecological degradation. Weak regulatory frameworks have enabled informal and illegal trade networks to flourish, intensifying the socio-environmental crises. Armed groups have also exploited the region's natural resources and waterways, making the watershed a focal point of conflict and resource extraction over the last four decades (Centro Sociojurídico para la Defensa Territorial Siembra 2022, 2024, González-Serrano 2024, Melo-Ascencio 2024). It is also clear that the involvement of illegal actors in both legal and illegal mining enterprises has granted them an outlet to both produce wealth and to move and launder this through national and international markets.

The region now faces an escalating environmental crisis characterized by inadequate waste management, poor potable water systems, ineffective sewage treatment, deforestation and large-scale illegal mining. These cumulative challenges threaten public health, food security and the overall resilience of the river's ecosystem and its communities, underscoring the urgent need for sustainable solutions (Urbam 2023).

The legal path to Judgment T-622 of 2016 was shaped by earlier judicial cases and advocacy efforts addressing environmental and human rights violations affecting the Atrato River and its surrounding communities. A key moment came in 2011 with a writ of constitutional protection (*acción popular*) filed for the Quito River, a tributary of the Atrato. This river had suffered severe environmental damage due to gold mining with heavy machinery, leading to deforestation, mercury contamination and ecosystem degradation. The case sought state accountability for environmental destruction and the failure to protect regional rivers, marking one of the first legal efforts to demand government action for ecosystems vital to Afro-Colombian and Indigenous communities (Gallón Droste 2021, 80). In 2015, another ruling on the Quito River reinforced the argument that state negligence had led to systemic rights violations in the Atrato Basin (Gallón Droste 2021, 80–81). The profound social and ecological harm caused by intensive extraction of natural resources in the Atrato River basin prompted the *Defensoría del Pueblo* (Ombudsman's Office) to assess conditions in Chocó in 2014. Its investigation exposed systemic failures, culminating in an official recognition of an environmental and humanitarian crisis in the region (Pèriz Fernàndez and González Moreno 2019, 21).

As Pèriz Fernàndez and González Moreno (2019) explain, the legal effort to protect the Atrato River began in 2014, spearheaded by Colombian NGO Tierra Digna as part of an educational and organizational programme to empower local communities. Through specialized training, ethnic-territorial leaders have learnt to use legal tools to defend their land and environment against pollution, deforestation and resource exploitation. Community leaders from Bajo Atrato have identified the river's severe environmental degradation – exacerbated by armed conflict and state neglect – as a critical threat. In 2015, the local organizations Foro Interétnico Solidaridad Chocó (FISCH) and Consejos Comunitarios Mayores del Atrato, with legal and technical support from Tierra Digna, filed a *tutela* against 24 state entities. These legal actions led to the 2016 Constitutional Court ruling which recognized the Atrato River as a subject of rights.

The Atrato River legal case

According to the plaintiffs, the Atrato River is more than a waterway; it is essential for survival, culture and identity. It provides food, water and transportation while serving as a space for social and cultural life. In Embera cosmology, the river plays a central role in creation myths, sacred rituals and healing practices. Women have an especially profound relationship with the river, using it for domestic tasks, community discussions and as a site of play and learning for children. As key plaintiffs in the *tutela*, women played a crucial role in documenting and articulating the river's significance, emphasizing their role as guardians of identity, memory and life (Pèriz Fernàndez and González Moreno 2019, 18).

The legal effort behind the lawsuit focused on framing the socio-environmental crisis of the Atrato River as a violation of multiple fundamental rights. Rather than limiting the case to the right to a healthy environment, the argument extended to violations of essential rights such as life, health, food, land, self-determination and free movement – affecting both collective entities (*consejos comunitarios* or Afro-Colombian community councils and *cabildos indígenas* or Indigenous councils) and the individuals who depend on the river for survival.

The case was developed in collaboration with four key community organizations: Ascoba (representing the Lower Atrato), FISCH (Foro Interétnico Solidaridad Chocó or Interethnic Solidarity Forum of Chocó),

Figure 3.2 The Atrato River winding its way through the department of Chocó. Photo by Daniel Alfonso León.

Cocomacia (representing the Middle Atrato) and Cocomopoca (representing the Upper Atrato). These groups played a central role in shaping the legal arguments, refining demands and collectively building the case. After a year of preparation, the *tutela* was ready to be filed. Although Indigenous groups did not directly participate in the litigation, their concerns were integrated through the FISCH platform. Ultimately, the case was led by these four Afro-Colombian organizations, reflecting a broader struggle for territorial and environmental protection in the region (Pèriz Fernàndez and González Moreno 2019, 19–20).

The petitioners in the Atrato case alleged a series of interconnected environmental and social harms driven by both informal and formal economic activities in the river basin. They highlighted the pervasive impact of informal mining, characterized by unregulated operations using mechanized equipment and toxic substances like mercury, which have caused widespread contamination of water sources and the destruction of aquatic ecosystems. Additionally, they pointed to the allocation of mining titles to corporations by the state, facilitating legal extraction activities that disregard the rights and wellbeing of local communities. The petitioners also underscored the issue of illegal deforestation driven by unauthorized logging, which degrades the region's forests and disrupts biodiversity. Corporate logging, carried out under state-sanctioned permits, was similarly criticized for contributing to deforestation while neglecting the ecological and cultural significance of the area. Furthermore, the discharge of untreated domestic and industrial waste into the Atrato River was identified as a significant source of pollution, further compromising the river's health and the ability of communities to depend on it for drinking water, fishing and agriculture (González-Serrano 2024).

While the petitioners invoked the violation of collective human rights to life, health, a healthy environment, water and the cultural rights of the Afro-Colombian and Indigenous communities living in the basin, they did not invoke the protection of the river's rights or biocultural rights. These concepts had no precedent in Colombian judicial practice and were not part of the community's claims in the case. This absence underscored the unprecedented nature of the court's ultimate decision, which went beyond the petitioners' claims to recognize the Atrato River as a subject of rights, expanding the scope of environmental and collective rights in Colombia.

The judicial process leading to the T-622 ruling was supported by a series of national and international norms, agreements and legal principles that validated the communities' demands. Drawing upon these legal frameworks, the Constitutional Court acknowledged the rights of the Atrato River and granted it personhood status, affirming its role as a subject of rights.

National laws like the Environment Act, Law 99 of 1993, which establishes Colombia's environmental protection framework, and the Afro-Colombian Territorial Rights Law 70 of 1993, recognizing Afro-Colombian territorial rights, were crucial to the case. Together with the 1991 Constitution, which guarantees the right to a healthy environment and requires the state to protect the country's ethnic and cultural diversity, these legal instruments created a favourable context for the claims brought forward by the communities of Chocó.

Tutela actions are initially filed before local judges, who have the authority to hear these claims when fundamental rights are alleged to have been violated by public authorities or, in certain cases, by private parties. These judges must process *tutelas* in a preferential and expedited manner, ensuring the immediate protection of the rights at stake. Once the local judge issues a ruling, the decision can be appealed before a higher court. Subsequently, the Constitutional Court has the power to review *tutela* rulings exceptionally, selecting cases that, due to their constitutional significance or social impact, warrant a definitive decision (Decree 2591, Government of Colombia 1991, art. 33).

The first-instance judges rejected the *tutela* for the Atrato River, arguing that the case did not meet the requirements for this type of legal action. They considered that the claims involved complex environmental and administrative issues that should be addressed through ordinary legal channels rather than a *tutela*, which is meant for the immediate protection of fundamental rights. Additionally, they determined that the case primarily concerned collective rights rather than individual rights, making an *acción popular* (similar to a class action) – a legal mechanism designed to protect collective interests such as environmental rights – the appropriate legal remedy (see Congress of the Republic of Colombia 1998). However, upon review, the Constitutional Court recognized the broader implications of the claims, including their direct impact on fundamental rights, and ultimately ruled in favour of the petitioners, granting rights to the petitioners and the river.

At the international level, treaties and principles such as the Rio Declaration on Environment and Development (1992) and the Convention on Biological Diversity (1993) reinforced the legitimacy of the communities' claims by recognizing the intrinsic connection between environmental protection and human rights. The Rio Declaration's principles of sustainable development and the precautionary approach underscored the state's duty to prevent environmental harm that endangers human wellbeing. Meanwhile, the Convention on Biological Diversity emphasized the protection of Indigenous and local communities' rights over their traditional territories and natural resources.

The T-622 ruling

The Court's decision to go beyond declaring a systematic violation of human rights in the Atrato ruling and to recognize the Atrato River as a subject of rights for the first time in the country was influenced by the process of granting legal personhood to the Whanganui River in Aotearoa New Zealand. As we learned through interviews, the Constitutional Court was particularly interested in making a structural pronouncement regarding the ecological crisis in the Atrato Basin. Initially, the Court considered issuing a structural ruling declaring only a breach of the constitution.

An "unconstitutional state of affairs" (*estado de cosas inconstitucional*) is a legal concept used by Colombia's Constitutional Court to address serious and recurring violations of fundamental rights affecting a specific population. This situation arises when these violations are widespread and result from systemic issues, rather than isolated incidents. It occurs when individual legal actions (such as *tutelas*) are insufficient to resolve the issue and broader, collective measures are needed. In such cases, the Court issues general directives to authorities to collaborate and resolve the problem. This mechanism aims to address widespread legal shortcomings and has been applied in various contexts, such as forced displacement, prison overcrowding and security concerns for former combatants (Uprimny 2022).

However, our sources indicated that the Court understood its administrative capacity to monitor structural rulings as increasingly exhausted, making it impossible to handle another structural case in the traditional manner. Inspired by the legal innovation in Aotearoa New Zealand and motivated to address the crisis in the Atrato Basin in a more structural way, the office of Magistrate Jorge Iván Palacio decided to introduce a novel legal approach. This approach recognized the structural rights crisis of the river itself, in conjunction with the human rights violations resulting from it, and designed mandates for compliance that involved various state entities at different levels, each with concrete obligations to address the basin's environmental and humanitarian crises. This framework aimed to ensure that these entities would work together to resolve the ongoing ecological and social challenges in the region, marking a significant shift in how environmental and human rights crises are approached in Colombian constitutional law.

The Constitutional Court issued Judgment T-622 in November 2016. Initially, a short version of the decision was released, followed by the full ruling with its complete justification in May 2017. In its ruling, the Constitutional Court explicitly acknowledged a turn towards an ecocentric paradigm as the foundation of the decision, challenging the traditional view of nature as a mere object for human use. Instead, it recognized nature as an entity worthy of legal protection and a subject of rights. This approach goes beyond utilitarian protection models. The Court introduced the concept of "biocultural rights" for the first time in Colombia, emphasizing the interdependence between nature and humans, especially for communities whose way of life is deeply connected to their traditional lands (Macpherson et al. 2020, Pèriz Fernàndez and González Moreno 2019, Sánchez-Zapata 2022).

Biocultural rights served as a legal framework to protect these communities' management of their land and resources, aiming to safeguard both biodiversity and cultural heritage. This category, according to the Court, seeks to ensure the participation of affected communities in public policy and legal frameworks, particularly in decisions impacting biodiversity conservation and cultural practices tied to land. The Court also highlighted widespread and systematic violations of human rights due to the neglect of various state institutions, outlining a path through its orders to protect the rights that were being severely violated (Pèriz Fernàndez and González Moreno 2019).

Further cases of rights of nature in Colombia

Inspired by the Atrato case, and perhaps driven by frustration over unresolved pollution issues, Colombia has increasingly turned to the rights of nature framework to enhance environmental protection. Since then, multiple ecosystems have been recognized as subjects of rights through both judicial rulings and legislative action. These decisions have extended legal personhood to key rivers, national parks and other ecological systems, mandating conservation, pollution control and restoration efforts. Notably, in 2024, Colombia took a legislative step forward with the approval of Law 2415, which declared the Ranchería River a subject of rights – marking a shift from judicial to legislative recognition. This law aims to ensure the river's protection and restoration, acknowledging its vital importance to Indigenous communities and regional biodiversity (Eco Jurisprudence Monitor 2025). A comprehensive overview of these cases, including the specific ecosystems recognized and their corresponding legal measures, is provided in Table 3.1.

Beyond these legal developments arising from the Atrato case, Colombia has also recognized nature as a *victim* of the country's long-running armed conflict. This means that environmental destruction is not just seen as collateral damage but as a direct consequence of violence, warranting legal redress. The Truth Commission, an official body created to investigate the impact of the conflict, has documented how war has caused severe environmental destruction, disrupting ecosystems and the deep connection that many communities have with their lands. Colombia's Special Jurisdiction for Peace (Jurisdicción Especial para la Paz – JEP), a transitional justice tribunal established under the 2016 peace agreement with guerrilla group Farc-EP, has taken steps to address this environmental harm. For instance, Order 079 of 2019 and Order SRVR 226 of 2023 specifically address the ecological damage caused by the conflict, paving the way for accountability and restoration efforts (JEP 2022).

Additionally, the Justice and Peace Tribunal (Tribunal de Justicia y Paz), which handled cases related to former paramilitary groups, has incorporated environmental harm into its framework. A key example is its 2016 ruling on the Magdalena River case, in which the tribunal acknowledged the extensive environmental destruction caused by paramilitary groups operating along Colombia's largest and most important river. The ruling detailed how illegal armed groups engaged in deforestation, contamination of water sources with toxic substances and destruction of wetlands and other ecosystems, severely affecting local communities that rely on the river for fishing, agriculture and cultural practices. By recognizing these environmental harms as part of the broader impacts of the armed conflict, the tribunal reinforced the idea that ecological degradation is not just an environmental issue but also a social one, directly affecting the livelihoods, health and cultural identity of affected populations and of nature itself (Eslava 2019).

Moreover, in 2019, Colombian legislators proposed a constitutional amendment to formally recognize nature's rights within Article 79, which already guarantees the human right to a healthy environment (Lozada Vargas et al. 2019). The proposal was not approved. However, it aligns with the broader international discourse promoted by the UN, which culminated in the 2022 global declaration on the human right to a healthy environment. Judicial decisions and law now encompass ecosystems that span approximately 72% of Colombia's territory, thereby recognizing these areas not merely as objects of administrative management but as legal subjects with participatory rights in governance (Sánchez-Zapata 2022). However, despite these advancements, implementation remains a significant challenge.

Although undoubtedly one of the most notable and frequently cited cases in the global literature on the rights of nature, the Atrato ruling was not the first of its kind in Colombia, at least not in terms of its philosophical foundation. One year before formally recognizing nature's rights in its 2016 Atrato River Basin ruling, the Colombian Constitutional Court advanced an ecocentric approach in a lawsuit concerning the protection of Tayrona National Park. In Judgment T-606/15, the Court emphasized that nature must be

Table 3.1. Rights of nature cases in Colombia. Table by authors.

Year	*Case*	*Court/Institution*	*Key Ruling & Significance*
2015	Tayrona National Park Case* (did not give legal subjectivity to the ecosystem)	Constitutional Court (T-606/15)	Ruled that nature must be protected for its intrinsic right to exist, not just for its instrumental value. Strengthened Indigenous consultation requirements.
2016	**Atrato River Case (*first case to afford legal personality to the ecosystem)**	**Constitutional Court (T-622/16)**	**Declared the Atrato River a subject of rights, mandating protection, conservation and restoration with appointed guardians.**
2018	Colombian Amazon Case	Supreme Court (STC4360–2018)	Recognized the Colombian Amazon as a subject of rights, ordering urgent action against deforestation and climate change.
2018	Páramo de Pisba Case	Superior Court of Boyacá	Declared the Páramo de Pisba a subject of rights, protecting it from mining and deforestation.
2019	Magdalena River Case	Administrative Tribunal of Cundinamarca	Declared the Magdalena River a subject of rights, recognizing its ecological and social significance.
2019	Cauca River Case	Superior Court of Medellín	Granted legal personhood to the Cauca River, emphasizing its cultural and ecological importance.
2019	Quindío River Case	Administrative Tribunal of Quindío	Recognized the Quindío River as a legal entity, ensuring its protection and restoration.
2019	Otún River Case	Fourth Court for the Execution of Sentences of Pereira	Declared the Otún River a subject of rights, mandating conservation and pollution control measures.
2019	Coello, Combeima and Cocora Rivers Case	Administrative Tribunal of Tolima	Recognized these three rivers as subjects of rights, requiring environmental preservation and restoration efforts.
2019	JEP – Order 079/2019 (Katsa Su Case)	Special Jurisdiction for Peace (JEP)	Recognized the Katsa Su (Awá Indigenous Territory) and 32 Indigenous councils as collective victims in Case 02, acknowledging environmental harm from the armed conflict.

Year	*Case*	*Court/Institution*	*Key Ruling & Significance*
2019	Pance River Case	Third Court for the Execution of Sentences of Cali	Granted Pance River legal rights, requiring governmental action to prevent contamination and promote restoration.
2019	La Plata River Case	Municipal Civil Court of La Plata-Huila	Recognized the La Plata River as a subject of rights, enforcing conservation and sustainable management.
2020	Vía Parque Isla de Salamanca Case	Supreme Court (STC3872–2020)	Recognized Vía Parque Isla de Salamanca as a subject of rights, ordering a strategic plan to halt deforestation and degradation.
2020	Lago de Tota Case	First Civil Court of Boyacá	Recognized Lago de Tota as a subject of rights, highlighting its importance for water resources and biodiversity.
2020	Páramo La Hermosa Case	Superior Court of the Judicial District of Ibagué	Recognized the Páramo La Hermosa as a subject of rights, reinforcing protections against climate change and human intervention.
2023	JEP – Order SRVR 226/2023 (Cauca River Case)	Special Jurisdiction for Peace (JEP)	Recognized the Cauca River as a victim of the armed conflict in Case 05, highlighting the need for environmental reparations and restoration.
2023	All-of-Nature Recognitions in Nariño and Boyacá	Various judicial and administrative bodies	Recognized all of nature as a subject of rights in Nariño and Boyacá, strengthening regional conservation efforts.
2024	Ranchería River Case	Colombian Congress (Law 2415)	First legislative recognition of a river as a subject of rights. Ensured protection and restoration, emphasizing its importance for Indigenous communities.
2024	Fortalecillas River Case	Administrative Tribunal of Huila	Declared the Fortalecillas River a subject of rights, requiring measures to prevent pollution and restore its ecosystem.

protected for its intrinsic right to exist, not merely for its instrumental or economic value. This ruling reinforced the importance of preserving ecosystems as legal entities and set a precedent for future environmental jurisprudence (Richardson and Bustos 2023).

Court-ordered remedial actions in the Atrato case

The Court's ruling required specific actions to mitigate the environmental impact of illegal mining. It mandated the government to design and implement a comprehensive decontamination strategy for the Atrato River basin, focusing on restoring affected areas, removing pollutants and reforesting degraded lands. This initiative aimed to rehabilitate the river's ecosystem while supporting communities dependent on the river's resources. Additionally, the Court ordered a plan to eradicate illegal mining operations in the Atrato basin, identifying them as a primary source of environmental harm. Government agencies were tasked with allocating resources to combat illegal mining, targeting both the river and surrounding areas within the Chocó department (Judgment T-622, 2016).

Recognizing the communities' reliance on the river for their livelihoods, the Court directed the development of culturally appropriate, sustainable programmes to revitalize traditional means of subsistence and ensure food security. It also required comprehensive studies on mercury contamination and other pollutants in the Atrato River and nearby communities. These studies aim to document the extent of contamination, assess health risks and guide future environmental protection policies (Judgment T-622, 2016, 158ff).

To enforce its orders, the Court established a monitoring and compliance framework with specific mechanisms to track implementation. A panel of experts was created to provide technical oversight and guidance, focusing on ensuring that the river's rights were upheld and that restoration measures followed scientific principles. The Court required government agencies to submit biannual compliance reports outlining progress on decontamination, enforcement against illegal mining and health outcomes. These reports were to be reviewed by the Constitutional Court and the Administrative Court of Cundinamarca to document actions taken and assess compliance with the ruling. To coordinate efforts across government bodies, the Court established the Inter-Institutional Commission for Chocó, tasked with overseeing the implementation of the ruling and addressing issues in the region. This commission should bring together agencies involved in environmental management, public health and social development to align their activities with the Court's orders (Judgment T-622, 2016, 158ff).

The Court assigned the Ministry of Finance and the National Planning Department the responsibility of securing and allocating financial resources to implement the measures outlined in the ruling. These agencies were instructed to provide the funding necessary for decontamination, enforcement and community support initiatives.

While the Constitutional Court's recognition of the Atrato River as a subject of rights and acknowledgement of human rights violations marked a legal milestone, it did not fully engage with all the petitioners' arguments. Ximena González-Serrano, one of the legal advisors who brought the case to court and now studies it academically, is a crucial voice on this topic. She has criticized the ruling for its failure to address critical issues surrounding state-licensed extractive activities, such as logging and legal mining. These activities are significant contributors to the river's degradation, and González-Serrano's expertise allows her to illuminate the broader implications of this oversight, advocating for a more comprehensive understanding of the environmental and legal ramifications of such practices. Though the decision established river rights, it avoided imposing binding obligations or remedies against extractive economies supported by state permits, concessions and policies. Critics argue that instead of resolving the legal and social conflicts caused by these pressures, the eco-jurisprudence framework expanded rights without assigning clear responsibilities to public or private actors driving economic policies (González-Serrano 2024, Melo-Ascencio 2024). Despite initial assessments identifying numerous environmental harm factors tied to extractive activities, institutional responses focused on issues like illegal logging and waste disposal, ignoring legal mining, industrial agriculture and proposed mega-projects (González-Serrano 2024, 13–14).

Envisioning the future: A community-centred blueprint for the Atrato River's biocultural rights

The vision for the future of the Atrato River was captured in an illustration created by one of our interviewees, an officer from Colombia's Contraloría General de la Nación (Office of the Comptroller General), as she talked to us in February 2023. As part of her oversight role, she monitors the implementation of the Atrato ruling, ensuring government accountability in protecting the river and supporting affected communities. With a career dedicated to working alongside local populations, she highlighted both the challenges and opportunities of fully realizing the river's biocultural rights.

In her illustration, a stilt house – a common structure in the Atrato basin – symbolizes Judgment T-622. The walls represent the ruling's four key mandates: decontaminating the river, eradicating illegal mining, securing food sovereignty and safeguarding human and environmental health. These pillars provide the foundation for protecting both the river and the communities that depend on it. The house itself remains stable only if the river's legally recognized rights – protection, conservation, maintenance and restoration – are upheld, ensuring ancestry, cultural preservation and territorial integrity for ethnic communities. Oversight by the panel of experts and the intersectoral commission is essential for enforcing these measures.

The roof represents the guardians of the Atrato River, who serve as stewards of the ruling. This group, composed of 14 community-appointed

Table 3.2. Summary of remedies in Decision T-622/16. Table by Cuesta Mosquera and Vallejo Piedrahíta (2024, 9).

Order/ remedy	*Content*	*Duty of compliance*	*Deadline for compliance*
1st & 2nd	Overturn fist instance and appeal rulings	—	None
3rd	Declaration of **systematic human rights violations** in the Atrato Basin due to state omissions.	Presidency of the Republic, several national institutions, regional governments of Chocó and Antioquia, several municipalities	None
4th	– Recognition of the **Atrato River as a legal entity** entitled to the rights to protection, conservation, maintenance and restoration. – Design and establish a commission of Guardians of the Atrato River	Presidency, Ministry of Environment and Communities	3 months to establish the commission
5th	Create and implement a **plan for river decontamination and restoration**	Ministry of Environment and Sustainable Development; Ministry of Finance and Public Credit; Ministry of National Defense; Codechocó (environmental authority of Chocó); Corpourabá (environmental authority of Urabá); Governments of Chocó and Antioquia provinces; 15 Municipalities in the Atrato River Basin; Communities	1 year
6th	Create a **plan to eradicate illegal mining** in Chocó	Ministry of National Defense; National Police; National Army; Attorney General's Office; Governments of Chocó and Antioquia; 15 Municipalities in the Atrato River Basin	6 months
7th	Create a **plan to revive traditional forms of subsistence and nutrition**	Ministry of Agriculture and Rural Development; Ministry of the Interior; Ministry of Finance and Public Credit; National Planning Department; Governments of Chocó and Antioquia; 15 Municipalities in the Atrato River Basin	6 months

Order/ remedy	*Content*	*Duty of compliance*	*Deadline for compliance*
8th	Conducting **toxicological and epidemiological studies**	Ministry of Environment and Sustainable Development; Ministry of Health; National Institute of Health; Codechocó; Corpourabá	3 months to start, 9 months for completion
9th	**Monitoring and support** mechanisms to ensure the fulfilment and execution of all orders	Office of the Attorney General; Ombudsman Office (*Defensoría del Pueblo*); Auditor General's Office	3 months to convene expert panel
10th	Create the "**Interinstitutional Commission for Chocó**"	Presidency of the Republic; National Government	1 month
11th	**Resources allocation** to ensure the fulfilment and execution of all orders	Ministry of Finance and Public Credit; National Planning Department	None
12th	Extend decision to ethnic communities in Chocó that are in a similar factual and legal situation as the plaintiffs.	—	None

guardians and one state-appointed guardian, plays a crucial role in translating biocultural rights into lasting, meaningful action. As one moves along the Atrato, the phrase "the Atrato River is life" resonates – a powerful reminder that achieving this vision requires institutional commitment and territorial governance reforms.

For biocultural rights to be effectively upheld, institutions must embrace them as a shared responsibility, rather than merely a legal obligation. This requires training state officials in ethnic and environmental rights, integrating the ruling into development plans at all levels of government and allocating dedicated budgets for implementation. Strengthening monitoring and compliance mechanisms is also critical to ensure long-term impact.

Beyond institutional efforts, community engagement and knowledge co-production are essential. Collaboration between community, institutional and academic actors can take the form of awareness campaigns, educational initiatives (e.g., radio programmes, podcasts, infographics and workshops) and research that integrates local knowledge. Studies on the river and the ruling should operate across three levels: (1) local governments and institutions conducting diagnostics, (2) community members and guardians engaging in public education and (3) research hubs developing knowledge from within the territory.

From a community-strengthening perspective, the principle of "safeguarding together" is central. Pedagogical initiatives, such as the creation of *Guardianitos* (little guardians) of the Atrato River – where children and youth learn to defend and care for the river – could instil long-term environmental commitment. Additionally, strengthening Indigenous and maroon guards in coordination with the Atrato guardians would enhance territorial governance and water governance.

Ultimately, the future of the Atrato River depends on a multi-layered approach that integrates institutional accountability, community empowerment and collaborative environmental stewardship. Only through these combined efforts can the vision of a living, protected and thriving Atrato become a reality.

The complex politics of turning mandates into reality

Over the eight years since the Atrato ruling was passed (at the time of writing), implementation efforts have focused on data collection and designing mandated plans aimed at river decontamination, eradicating illegal mining and restoring traditional community livelihoods and food security. The state institution Codechocó has emphasized the critical importance of this preparatory phase, cautioning that without a strong foundation interventions could worsen environmental issues, such as shifting mercury contamination from sediments to the water. The construction of these action plans has been described as a participatory process involving significant contributions from the guardians of the Atrato River and local communities:

> Communities will not be mere passive actors. The ruling established mechanisms for citizen participation, which means that public entities will not unilaterally decide what actions to take. Instead, the orders specify that an action plan must be developed – with whom? With organizations and local people. For what purpose? To ensure that institutional decisions align with the realities of the communities that live there. This approach guarantees meaningful participation. Communities need to be aware of the proposed solutions to the social, environmental and economic challenges they face. Under the framework of the Fifth mandate order, we worked for nearly three years on the formulation of an action plan. Throughout this process, we engaged in continuous validation with the communities, recognizing their deep knowledge of the territory. Many of their observations – whether from local experts or community members themselves – were incorporated into the final document. Participation was not limited to individuals without formal education; there was also an advisory committee that included the Interethnic Solidarity Fund of Chocó, the Guardians of the Atrato River and a technical advisor from other organizations. The ultimate goal was clear: we were all working toward the same objective.
>
> (Personal interview with a representative from the Sub-Directorate of Quality and Environmental Control, Codechocó, February 2023)

Our interviews with Indigenous and Afro-descendant community members from the Atrato River, conducted in February 2023, reveal that riverside communities generally view the ruling as a meaningful contribution to their long-standing political struggles, despite significant challenges in its implementation. For them, the decision not only acknowledges the region's enduring humanitarian crises but also presents an opportunity to address critical issues in the Chocó department, particularly the devastating effects of illegal mining. Additionally, the ruling has helped legitimize and defend Afro-descendant artisanal mining practices, reducing stigmatization and reinforcing their cultural and economic significance. Still, the ruling has seen minimal progress towards its goal of halting illegal mining and decontaminating the river. Expanding ethnic community participation in consultations and decision-making related to its implementation also remains a critical area for improvement (Cuesta Mosquera and Vallejo Piedrahíta 2024).

As highlighted earlier, the Atrato ruling introduced an ambitious set of mandates for government ministries and public institutions. The Constitutional Court established short timelines for these mandates to set high aspirations and compel institutions to act with urgency. While these deadlines were recognized as practically unachievable, they sought to counteract institutional tendencies to indefinitely delay substantive interventions. The Court's ambitious goals aimed to drive immediate and tangible action.

However, eight years after the ruling, significant challenges persist. Field research conducted in Quibdó and Bogotá during 2022 and 2023 highlights numerous obstacles to fulfilling the ruling's mandates. These challenges, described as part of a "muddy political ecology" (McNeish and Moreno Socha 2024), include weak institutional coordination, inadequate funding, limited public consultation, insufficient legal education and ongoing environmental contamination and violence linked to illegal gold mining.

Despite its global significance as an innovative example of environmental law-making – frequently cited in cases concerning the rights of nature – the implementation of the Atrato ruling has been hindered by systemic issues. Representatives of regional Indigenous organizations reported insufficient consultation during the ruling's development and inadequate representation in the mandated guardianship model, where only two of the 14 guardians are Indigenous. Additionally, civil society organizations, river guardians and local government representatives have highlighted the lack of dedicated public funding. Guardians are left to finance their responsibilities, including extensive travel across the 45,000 km^2 watershed, without government support.

Stakeholders from non-governmental organizations at local, regional and national levels have also emphasized poor coordination among government ministries and institutions tasked with implementing the ruling. While the COVID-19 pandemic partly contributed to delays, many interviewees attributed inaction to flawed coordination and conflicting interpretations of the mandates. Members of the expert panel noted that while the ruling introduced the concept of biocultural rights, there is no consensus on its definition or practical application.

Most concerning is the limited progress in protecting the Atrato River and addressing illegal mining. Toxicological studies have revealed alarming levels of heavy metals in the riverbanks and in the blood and hair of local populations. Yet, there is no consensus on a remediation strategy. Despite ongoing efforts by the Ministry of Defence to raid mining camps and destroy equipment, illegal mining remains widespread. Observations in Quibdó reveal the open sale of gold and mining equipment, alongside continued violence from illegal armed groups dominating the region's economy.

The National Development Plan of the current Gustavo Petro government outlines a vision of territorial planning that recognizes water as a fundamental right, a common good and a core element of governance. This approach includes protections for oceans, coastal areas and islands, as well as the recognition of the communities living in these regions (Government of Colombia 2022, 59). The current administration's policies align with the Constitutional Court's Judgment T-622/16.

But while President Gustavo Petro's administration (2022–2026) is prioritizing environmental policies, it faces similar political constraints as previous governments in addressing the interconnected challenges of illegal mining and the drug trade. The signing of the Escazú Agreement favours a more positive outlook for environmental participation, and the National Development Plan

(NDP) promises increased funding for the Chocó region. Still, the funding specifically allocated for implementing the Atrato ruling remains insufficient. Eight years after the ruling, contamination, violence and insecurity surrounding the Atrato River have worsened rather than improved (Robayo Corredor et al. 2023).

Conclusions

The Atrato River case marks a shift in Colombia's legal landscape towards the recognition of nature's rights. Constitutional Court Judgment T-622/2016 established the Atrato River as a legal subject with four key rights: protection, conservation, maintenance and restoration. The ruling reflects a broader shift from traditional environmental protection frameworks towards ecocentric models, recognizing the interconnectedness of humans and ecosystems. This legal recognition aligns with a growing global trend of acknowledging the intrinsic rights of nature, which has gained increasing support within legal systems worldwide. Colombia is at the forefront of this trend.

Despite the legal advances, the implementation of the ruling has encountered significant challenges. These include weak institutional coordination, insufficient funding, continued environmental contamination and the persistence of illegal mining. Moreover, the lack of meaningful public consultation and inadequate representation of Indigenous communities in the guardianship model are critical issues that have hindered the effective realization of the ruling's provisions. The persistence of extractive industries, such as state-licensed logging and mining, and the lack of water and sanitation infrastructure further complicate the situation. While the ruling addresses the river's biocultural rights, the broader environmental, extractive and infrastructural issues highlighted by petitioners remain insufficiently addressed.

The challenges in implementing the Atrato ruling underscore the complexity of translating legal recognition into tangible environmental and social outcomes. While current government initiatives such as the National Development Plan and the ratification of the Escazú Agreement aim to improve environmental governance, the limited financial resources allocated for the Atrato ruling's implementation remain a significant barrier. As a result, environmental contamination, violence and insecurity in the region continue to worsen, reflecting the difficulties inherent in securing long-term, meaningful change.

Moving forward, the effective implementation of the Atrato ruling will require continued collaboration among government bodies, civil society and the organizations representing the region's collective territories. This collaboration must focus on overcoming systemic barriers such as institutional fragmentation, lack of resources and the persistence of extractivism. The successful realization of the river's rights depends on the creation of effective governance frameworks that integrate community participation and foster the protection of both the river and the surrounding community and their cultural and environmental rights.

In conclusion, the Atrato ruling offers an initial and crucial step towards advancing environmental justice and the recognition of nature's rights, but its successful implementation hinges on addressing the systemic challenges that continue to impede progress. The case provided a model for legal frameworks that prioritize the interrelated rights of nature and humans, but it also highlights the need for comprehensive, integrated approaches to environmental governance. The recognition of the river's rights is a first step, but it is only through sustained efforts and systemic changes that these rights can be fully realized.

Notes

1 The right to collective ownership of the land was earned by Indigenous and Black communities through decades of political mobilization, culminating in its formal recognition in the 1991 Constitution of Colombia. It was further developed and detailed in Law 70 of 1993, which reinforced the rights of these communities to collective land ownership and governance. See Law 70 (Congress of the Republic of Colombia 1993); see further Erazo Gómez (2012); Government of Colombia (2025).

2 "Indian reduction" was a colonial policy implemented primarily by the Spanish Crown in the Americas during the 16th and 17th centuries. Its goal was to concentrate dispersed Indigenous populations into settlements or communities under colonial control, making their evangelization, taxation and labour exploitation more efficient. See further Jurado (2004); Luna and Alburqueque (2020).

References

Aguirre Ceballos, Jaime and Jesús Orlando Rangel. 2008. "El Chocó biogeográfico." In *Colombia diversidad biótica VI: Riqueza y diversidad de los musgos y líquenes en Colombia*, edited by Jesús Orlando Rangel, 77–84. Universidad Nacional de Colombia.

Almario García, Óscar, Luis Javier Ortiz Mesa and Lina Marcela González Gómez. 2015. *El Chocó en el siglo XIX: Encrucijada histórica, social, territorial y conceptual. Hacia un nuevo siglo XIX del noroccidente colombiano.* Balance bibliográfico de Antioquia, Caldas y Chocó. Vol. 3. Universidad Nacional de Colombia.

Bonza, Luis. 2024. "Servidores públicos y Policía: cómplices de la minería ilegal en el Chocó." *Voragine*, 27 October 2024. https://voragine.co/historias/reportaje/servidores-publicos-y-policia-complices-de-la-mineria-ilegal-en-el-choco/.

Centro Sociojurídico para la Defensa Territorial Siembra. 2022. *El Atrato es la vida: Conflicto armado, economías extractivas y resistencias pacíficas en el Atrato.* https://www.youtube.com/watch?v=vtyXw4kLH7o.

Centro Sociojurídico para la Defensa Territorial Siembra. 2024. "¿Por qué ha fracasado la estrategia militarista contra la minería ilegal? El caso de La Cuenca Del Atrato Bogotá." https://www.centrosiembra.org/2024/04/18/por-que-ha-fracasado-la-estrategia-militarista-contra-la-mineria-ilegal-el-caso-de-la-cuenca-del-atrato/.

Congress of the Republic of Colombia. 1993. "Ley 70 de 1993, por la cual se desarrollan los derechos de las comunidades negras y el artículo transitorio 55 de la Constitución Política." https://www.funcionpublica.gov.co/eva/gestornormativo/norma.php?i=7388.

Congress of the Republic of Colombia. 1998. "Law 472 of 1998 by which Article 88 of the Political Constitution of Colombia is developed with respect to the exercise of popular and group actions and other provisions are enacted." https://www.funcionpublica.gov.co/eva/gestornormativo/norma.php?i=188.

Constituent Assembly of Colombia. 1991. *Political constitution of Colombia* (English trans. A.I. Vellvé Torras & J.J. Ruchti). *Banco de la República*. https://www.banrep.gov.co/en/regulation/political-constitution-colombia-1991.

Constitutional Court of Colombia. 2016. Judgment T-622. Atrato River Case.

Cuesta Mosquera, Arlinton, and Catalina Vallejo Piedrahíta. 2024. "Take me to the River: Have riverine rights enhanced community participation in environmental governance within the Atrato River Basin?" *The International Journal of Human Rights* (July): 1–19. doi:10.1080/13642987.2024.2367468.

Eco Jurisprudence Monitor. 2025. https://ecojurisprudence.org/.

Erazo Gómez, Alejandra. 2012. "La Ley 70 de 1993 y la participación política de las comunidades negras." *Trans-pasando Fronteras* 2: Article 2. doi:10.18046/retf.i2.1321.

Eslava, Gabriela. 2019. "Naturaleza: ¿víctima del conflicto?" *Dejusticia*, February 8. https://www.dejusticia.org/naturaleza-victima-del-conflicto/.

Gallón Droste, Elizabeth. 2021. "Navegar futuros por ríos de oro." *Protagonists of Latin American Futures* 9 (1): 75–85.

García Reyes, Paola. 2014. "Tierra, palma africana y conflicto armado en el Bajo Atrato chocoano, Colombia: Una lectura desde el cambio en los órdenes de extracción." *Estudios Socio-Jurídicos* 16 (1): 209–244. doi:10.12804/esj16.1.2014.05.

Gobernación del Chocó. 2020. *Plan de desarrollo departamental 2020–2023 Chocó: Generando confianza*. https://rap-pacifico.gov.co/wp-content/uploads/2020/07/choco-final-pddd.pdf.

González-Serrano, María Ximena. 2024. "Rights of nature, an ornamental legal framework: Water extractivism and backbone rivers with rights in Colombia." *The Journal of Peasant Studies* 0 (0): 1–21. doi:10.1080/03066150.2024.2349228.

Government of Colombia. 1991. "Decree 2591." https://www.funcionpublica.gov.co/eva/gestornormativo/norma.php?i=5304.

Government of Colombia. 2022. "Plan Nacional de Desarrollo 2022–2026." https://www.dnp.gov.co:443/plan-nacional-desarrollo/pnd-2022-2026.

Government of Colombia. 2025. "Comunidades negras, afrocolombianas, raizales y palenqueras." https://www.unidadvictimas.gov.co/comunidades-negras-afrocolombianas-raizales-y-palenqueras/.

HRC (United Nations Human Rights Committee). 1984. "General Comment No. 12: The Right to Self-Determination of Peoples (Article 1, ICCPR)." March 13. https://www.refworld.org/docid/4538837e198.html.

IACtHR (Inter-American Court of Human Rights). 2007. "Saramaka People v. Suriname. Judgment of November 28, 2007." https://www.corteidh.or.cr/docs/casos/articulos/seriec_172_ing.pdf.

IACtHR (Inter-American Court of Human Rights). 2015. "Case of the Kaliña and Lokono Peoples v. Suriname. Judgment of November 25, 2015." https://www.corteidh.or.cr/docs/casos/articulos/seriec_309_ing.pdf.

ICJ (International Court of Justice). 1975. "Western Sahara, Advisory Opinion." ICJ Reports. https://www.icj-cij.org/en/case/61/advisory-opinions.

ILO (International Labour Organization). 1989. "Convention No. 169 concerning Indigenous and Tribal Peoples in Independent Countries. June 27, 1989." https://

www.ilo.org/dyn/normlex/en/f?p=NORMLEXPUB:12100:0::NO::P12100_INSTRUMENT_ID:312314.

JEP (Jurisdicción Especial para la Paz). 2022. "El ambiente como víctima silenciosa." https://www.jep.gov.co/JEP/documents1/El%20ambiente%20como%20v%C3%ADctima%20silenciosa.pdf.

Jiménez Meneses, Orián. 2001. "La provincia del Chocó ante el 'Estado Nación'." *Anuario de Historia Regional y de las Fronteras* 6 (1): 92–119.

Jurado, Carolina. 2004. "Las reducciones toledanas a pueblos de indios: Aproximación a un conflicto." *Cahiers des Amériques latines* 47. doi:10.4000/cal.7814.

Kerguelén Sánchez, María Claudia, Roberth Paternina-Uribe, Elvia Valeria Durante-Yánez, María Alejandra Martínez-Macea and José Luis Marrugo-Negrete. 2022. "Capítulo 1. Exposición humana al mercurio en una población al norte de Colombia, cuenca del río Atrato." In *Atrato y salud: Resultados del proyecto 849-2018*, edited by José Luis Marrugo-Negrete, Roberth Paternina-Uribe, Manuel Haminton Salas-Moreno and Carlos Adolfo Salazar-Camacho, 13–46. Universidad de Córdoba.

Leal León, Claudia María. 2016. "Freedom in the rainforest: The formation of a black peasantry in Colombia's Pacific Coast, 1850–1930." *Revista CS* 20, Article 20. doi:10.18046/recs.i20.1861.

Lozada Vargas, Juan Carlos, Harry Giovanny González García, Juan Fernando Reyes Kuri, Andrés David Calle Aguas, Alejandro Alberto Vega Pérez, Nilton Córdoba Manyoma, José Daniel López Jiménez, Inti Raúl Asprilla Reyes, Julián Peinado Ramírez and Juan Carlos Rivera Peña. 2019. *Por el cual se modifica el artículo 79 de la Constitución Política de Colombia*. Cámara de Representantes. https://www.camara.gov.co/derechos-de-la-naturaleza.

Luna, Gillermo Figueroa, and Iris Liliana Vásquez Alburqueque. 2020. "Dos procesos olvidados: La catástrofe demográfica y la imposición de 'reducciones de indios' en el norte peruano." *Investigaciones sociales* 23 (43): 239–249.

Macpherson, Elizabeth, Julia Torres Ventura and Felipe Clavijo Ospina. 2020. "Constitutional law, ecosystems, and indigenous peoples in Colombia: Biocultural rights and legal subjects." *Transnational Environmental Law* 9 (3): 521–540. doi:10.1017/S204710252000014X.

McNeish, John-Andrew, and Jenny Moreno Socha. 2024. "Muddy waters: On the problematic political ecology of the Atrato ruling, Colombia." *The International Journal of Human Rights* 0 (0): 1–21. doi:10.1080/13642987.2024.2324121.

Melo-Ascencio, Diego Alejandro. 2024. "A postcolonial Marxist critique of the river-as-subject: Situating the Atrato River in its development aporias." *Antipode* 56 (5): 1754–1774. doi:10.1111/anti.13052.

OAS (Organization of American States). 2016. "American Declaration on the Rights of Indigenous Peoples (ADRIP)." June 15. https://www.oas.org/en/sare/documents/DecAmIND.pdf.

OAU (Organization of African Unity). 1981. "African Charter on Human and Peoples' Rights (Banjul Charter)." June 27. https://au.int/en/treaties/african-charter-human-and-peoples-rights.

Palacios-Torres, Yuber, Karina Caballero-Gallardo and Jesus Olivero-Verbel. 2018. "Mercury pollution by gold mining in a global biodiversity hotspot, the Choco Biogeographic Region, Colombia." *Chemosphere* 193 (February): 421–430. doi:10.1016/j.chemosphere.2017.10.160.

Pérez-Escobar, Oscar Alejandro, Eve Lucas, Carlos Jaramillo, Alexandre Monro, Sarah K. Morris, Diego Bogarín, Deborah Greer, Steven Dodsworth, José Aguilar-Cano and Andrea Sanchez Meseguer. 2019. "The origin and diversification of the hyperdiverse flora in the Chocó Biogeographic Region." *Frontiers in Plant Science* 10: 1328.

Pèriz Fernàndez, Elisabet, and Viviana González Moreno. 2019. "Defendiendo el río Atrato: reflexiones del caso y apuntes sobre el rol de las mujeres en el proceso de defensa del territorio." *UN Women Colombia*. https://colombia.unwomen.org/sites/default/files/Field%20Office%20Colombia/Documentos/Publicaciones/2019/12/3%20Caso%20Sentencia%20rio%20Atrato.pdf.

Poveda, Ingrid Carolina, Camilo Andrés Rojas, Agustín Rudas and J. Orlando Rangel. 2004. "El Chocó biogeográfico: Ambiente físico." *Colombia diversidad biótica IV, El Chocó biogeográfico/Costa Pacífica*: 1–22.

Richardson, Whitney, and Camila Bustos. 2023. "Implementing nature's rights in Colombia: The Atrato and Amazon experiences." *Revista Derecho Del Estado* 54: 227–275.

Robayo Corredor, Alejandro, Allison Kielhold, Ana María Rodríguez Contreras, Angela Ramirez Rincon, Brenda Forero Linares, Carolina Serrano Idrovo, Cristian Saez Florez, et al. 2023. "Hacia una paz sostenible: Un análisis de la implementación del acuerdo final y su relación con el ambiente." Report. University of Notre Dame, Kroc Institute, August 4. doi:10.7274/24870246.v1.

Rodríguez Hernández, Nelson Eduardo. 2014. "Cartografía de la frontera 'bárbara': Las representaciones del Darién a propósito del conflicto entre el Virreinato de Nueva Granada y los Cunas." *Anuario de Historia Regional y de las Fronteras* 19 (1): Article 1.

Sánchez-Zapata, Diana Carolina. 2022. "La naturaleza como sujeto de derechos: Un análisis de su incidencia en la transformación de la planeación del ordenamiento territorial en Colombia desde un enfoque ecocéntrico de sostenibilidad ambiental fuerte y de los derechos bioculturales." PhD thesis. Universidad Nacional de Colombia, Sede Medellín.

Sharp, William Frederick. 1976. *Slavery on the Spanish Frontier: The Colombian Chocó 1680–1810*. https://repositorio.ucaldas.edu.co/handle/ucaldas/17448.

Tubb, Daniel. 2020. *Shifting livelihoods: Gold mining and subsistence in the Chocó, Colombia*. University of Washington Press.

Universidad de Córdoba and Universidad Tecnológica del Chocó. 2022. *Elementos traza en el atrato: El mercurio no es el único problema. IV Seminario Internacional de Ciencias Ambientales SUE-Caribe* [Video]. https://www.youtube.com/watch?v=VCHL6ikSDLM.

Uprimny, Rodrigo. 2022. "5 preguntas para entender el estado de cosas inconstitucional por la inseguridad de excombatientes." *Dejusticia*. https://www.dejusticia.org/5-preguntas-para-entender-el-estado-de-cosas-inconstitucional-por-la-inseguridad-de-excombatientes/.

Urbam. 2023. *Hoja de Ruta del Atrato: Plan Estratégico del Río Atrato para los Municipios de Murindó, Vigía del Fuerte, Turbo y Mutatá, Tomo I Análisis Estratégico contextual*. Gobernación de Antioquia; Universidad EAFIT; Centro de Estudios Urbanos y Ambientales Urbam. https://bibliotecadigital.udea.edu.co/entities/publication/d967d227-0a5d-40ed-9e42-f79a5a054029.

Vargas, Patricia. 1990. "Los emberas y los cunas en frontera con el imperio español: Una propuesta para el trabajo complementario de la historia oral y de la historia documental." *Boletín Museo del Oro* 29: Article 29.

4 India's Ganga and Yamuna rivers as legal persons

Axel Borchgrevink and Kavita Upadhyay

Introduction

In March 2017, the Uttarakhand High Court in India passed two orders that garnered public attention. In the first order, passed on March 20, India's national river Ganga (Ganges) and another river, Yamuna, which is the Ganga's longest tributary, were declared living entities with all the rights, duties and liabilities of a legal person (Mohd. Salim v. State of Uttarakhand & Others 2017). The order was yet another way in which attention was brought back to the deteriorating health of the two inter-state rivers, especially the Ganga, which supports the lives of more than 600 million people (World Bank 2015) and yet has been battling for survival against the pressures of pollution and development. While the idea for legal rights of rivers introduced by the order was still fresh and lacked clarity, especially with regards to its implementation, another order was passed by the same high court on March 30 that bestowed legal personhood on all of nature (including glaciers, rivers, jungles, meadows and wetlands) in Uttarakhand (Lalit Miglani v. State of Uttarakhand & Others 2017). This order aimed at conservation of nature within Uttarakhand through the rights of nature approach. However, both the orders were challenged in the Supreme Court of India and remain pending.

While the two orders represented a significant break from the prevailing legal norms in India, the question of how exactly this shift would make way for cleaner and healthier rivers and nature remained, since the two orders lacked clarity in their view of rights of rivers and rights of nature approaches. In this chapter, we explore the reasons that led to the two orders, the gaps in the orders and the impediments to their implementation. We also delve into the religious sentiments the orders invoke, especially in relation to the Ganga – a river of great cultural and religious significance, considered sacred by the Hindus. What the orders may mean for other religious communities, and the marginalized, whose concerns they do not address, is discussed. Despite the existing gaps, the positive impacts of the orders are discussed too.

For the analysis, we use the Public Interest Litigations (PILs) filed by Mohd Salim (Mohd. Salim v. State of Uttarakhand & Others 2014) and Lalit

DOI: 10.4324/9781003581208-4

Miglani (Lalit Miglani v. State of Uttarakhand & Others 2015), which form the basis for the two orders that bestowed legal personality status on the Ganga and Yamuna rivers, and nature in Uttarakhand. The orders passed and judgments delivered in relation to both PILs form the basis for the analysis. The information gathered from interviews with several lawyers in the Uttarakhand High Court, including those directly involved in these PILs, also guided us in the directions explored in this chapter.

In the following sections, we provide the context for the chapter, followed by a brief history of environmental law and PILs in India, and a discussion of the rights of rivers and rights of nature orders in view of the PILs that led to their issuance. A background on the status of the Ganga, particularly regarding pollution, mining and infrastructure projects impacting the river, along with various civil society initiatives aimed at addressing these issues, is mentioned. The role of Justice Rajiv Sharma, who authored the two orders, is examined, especially in the context of judicial activism. This is followed by a discussion of whether Justice Sharma's act of breaking from the existing jurisprudence served any purpose, whether the rights of rivers and rights of nature orders were even implementable, and what impact the two orders could have on the rights of the marginalized population. We conclude with examining critical issues in relation to the notion of causing harm to nature that are mentioned in the court rulings – what is considered as harming nature, and what makes a case for prosecuting the acts of harming nature in the rights of rivers and rights of nature orders. In the concluding section, we also delve into the impact, even if indirect, of the two orders in paving the way for rights of nature through the Indian legal system.

Context

The Ganga originates at Gaumukh, the snout of the Gangotri Glacier, in Uttarakhand – a state in the Indian Himalayan Region that shares an international border with China and Nepal. Flowing for over 2,500 km, the river traverses other Indian states of Uttar Pradesh, Bihar, Jharkhand and West Bengal, and being a transboundary river flows thorough Bangladesh too. The Yamuna is the Ganga's longest tributary and flows through Uttarakhand – originating at the Yamunotri glacier – and other Indian states including Himachal Pradesh, Haryana, Uttar Pradesh and the National Capital Territory of Delhi. The confluence of the Ganga and the Yamuna is at Prayagraj in Uttar Pradesh.

The Ganga holds great religious and economic significance. For Hindus, the Ganga is a sacred river. Her water, called *Gangajal* in Hindi, is used in religious rituals, and people living far from the river may have bottles of it. It is also the national river of India, with great symbolic importance (Brandshaug et al. this volume, Borchgrevink this volume). Many hydropower projects have been built on the river, and mining of riverbed material is rampant in the Ganga. The river, however, is highly contaminated due to untreated

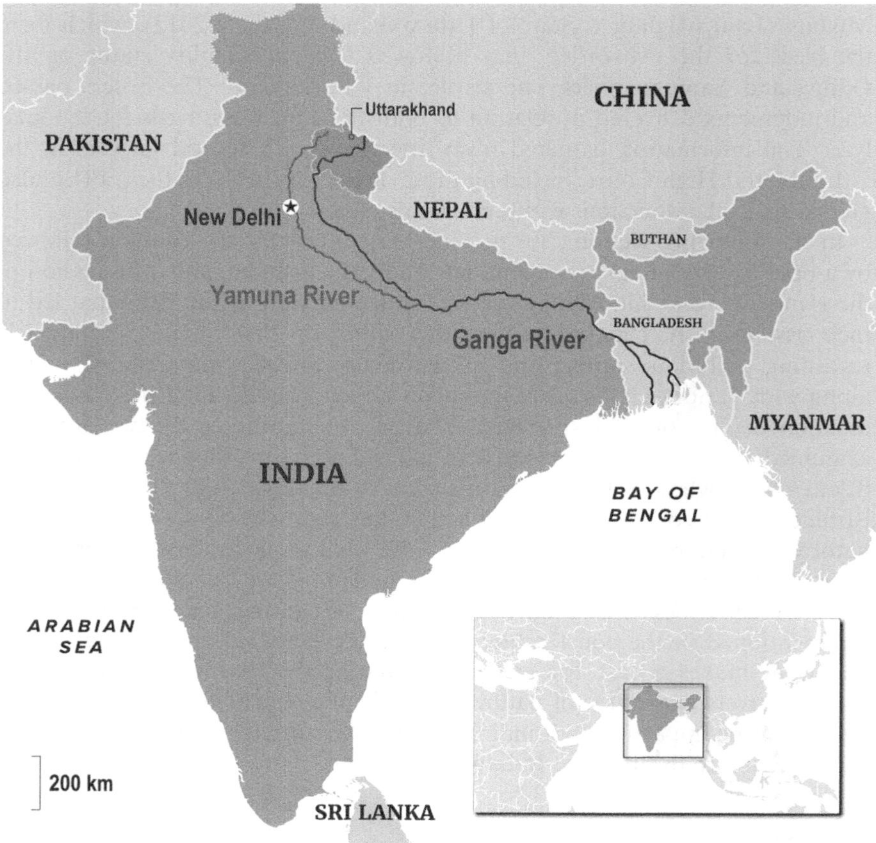

Figure 4.1. The Ganga and Yamuna rivers. Map data ©2025 Google. Elaborated by Daniel Alfonso León.

sewage and industrial waste being released into her. While various government initiatives, including the Namami Gange Programme, launched by the Indian government in 2014, have aimed at cleaning the Ganga and enforcing norms to reduce the discharge of untreated wastewater in the river, the Ganga continues to battle for her existence.

In Uttarakhand, the construction of infrastructure, like roads along rivers and hydropower projects (several of which are on the Ganga's tributaries), also impacts rivers, as muck is illegally dumped into and beside them during construction to save costs (Upadhyay 2024). The muck dumped beside rivers flows into them during extreme rainfall events, which are increasing due to climate change (IPCC 2021). The high sediment load in the river due to this muck increases the risk of flooding (Upadhyay 2024).

Activists, the civil society and religious organizations have been campaigning for a clean and healthy Ganga for decades (Borchgrevink this volume). In this regard, the journeys of two individuals are especially compelling – Swami Nigamananda Saraswati and Professor GD Agarwal, later known as Swami Gyan Swaroop Sanand. Both lost their lives while advocating for government action to preserve the river's sanctity and purity. Swami Nigamananda died in June 2011 after undertaking an indefinite fast to stop illegal mining in the river. It is alleged that he died from poisoning (Tripathi 2011). Professor Agarwal observed several fasts between 2008 and 2018 and died in 2018 while fasting with demands that included the passing of the Ganga Protection Act, a ban on sand mining in the Ganga in Uttarakhand's Haridwar, and scrapping all proposed and under-construction hydropower projects on the river (Upadhyay 2018). However, despite a long history of civil society campaigns for restoring the Ganga, there has been little interest in demanding legal personhood or rights for the river.

Public Interest Litigation and environmental law in India

Environmental legislation has a relatively short history in India. The Constitution, which came into effect in 1950, is widely credited for laying the foundations for a secular state with concerns for social protection. However, environmental issues were not included. The Water (Prevention and Control of Pollution) Act was passed in 1974 due to widespread concern over contamination of rivers and water sources. It was meant to give states the instruments for regulating contamination (Macpherson et al. 2021). In 1976, environmental protection was taken into the Constitution as a responsibility of the State (Article 48A). Moreover, the new Article 51A (g) stated that "It shall be the duty of every citizen of India to protect and improve the natural environment including forests, lakes, rivers and wildlife and to have compassion for living creatures". Implementation of these new laws has however been weak and inconsistent.

Since the late 1970s, Indian courts have been quite liberal in opening for PIL. With the explicit objective of giving access to justice for people who due to poverty or other limitations have traditionally been denied this, the courts have developed a procedure that is fairly unique to India (Chandra 2018, Holladay 2012). Both the Supreme Court of India and the high courts at state level have been willing to accept cases deemed to have a public interest, without strict criteria of standing or formal requirements of case presentation and sometimes with court-paid lawyers assisting petitioners in gathering relevant facts.

Indian courts have been important for developing legislation, also in the environmental field. Through PILs, The Indian Supreme Court has been important for the development of Indian constitutionalism, including what is often referred to as environmental constitutionalism (Macpherson et al. 2021, 441–438). Through a series of cases, it has expanded the reach of the constitutional "right to life and liberty" to include, among other things, the right to food, health, dignity, education, clean air and water and freedom from environmental pollution (Macpherson et al. 2021, 444).[1] One particular figure who has extensively used the PIL for environmental purposes is the lawyer MC Mehta, nicknamed "The Green Avenger". He initiated his activist career in the 1980s by filing a case in the Supreme Court to stop the contamination of the Ganga, which after several years ended in the closure of hundreds of tanneries and the introduction of new regulatory standards. Among his large number of subsequent PILs, many have dealt with issues of the Ganga.

The cases in the Uttarakhand High Court

The court cases that led to the personhood decisions in Uttarakhand High Court were also PILs: Mohd. Salim v. State of Uttarakhand & Others (2014), and Lalit Miglani v. State of Uttarakhand & Others (2015). The two cases were both long-running, and the processes were to some extent intertwined, including being decided by the same judges. It is therefore most convenient to present them in parallel. As the presentations also show, it is not far-fetched to view the decisions in these cases as examples of what Chandra has characterized as judges using their "unchecked power in capricious ways" (2018, 712).

Mohd Salim filed his case on July 30, 2014, objecting to illegal mining in the Ganga's riverbed and to unauthorized construction along the river by people from the neighbouring state of Uttar Pradesh. At a later point, the claim was expanded to also object to industrial waste being dumped illegally in the river. According to Nakazora, the encroachments he asked to have removed included the illegal slum settlements of migrant workers along the river. Mohd Salim had established a tourist resort "village" on the river and was concerned about the "spoilt landscape" (Nakazora 2023, 180–181). Hearings were held through 2014, 2015 and much of 2016 without any important order being passed. In September 2016, Justice Sharma (now retired) became a Judge in the Uttarakhand High Court, and on December 5, the same year, the court delivered a first judgment on the case, written by Justice Sharma (Mohd. Salim v. State of Uttarakhand & Others 2016). It required the eviction of the encroachers, a ban on mining, the settlement of the property dispute between Uttarakhand and Uttar Pradesh, the establishment of a Ganga Management Board, and Uttarakhand to be named on the Upper Yamuna River Board.

The other case had been filed on August 19, 2015, by Lalit Miglani, a practising lawyer at the Uttarakhand High Court. It concerned the lack of government action in response to the contamination of the Ganga due to the release of industrial waste and untreated sewage. It likewise had hearings in 2015 and 2016, before a judgment was passed on December 2 (three days before the decision in the Mohd Salim case mentioned above). This 79-page verdict went into details about the scientific data on the conditions of the river. In conclusion, orders to control and reverse contamination were given, as well as general prohibitions on defecation within 500 meters of the river and on the washing of cattle in the river (Lalit Miglani v. State of Uttarakhand & Others 2016).

It is worth pointing out that the directives of both these decisions are comprehensive and do not appear to be easily implementable in the short term. However, due to the non-implementation of the directives of the December 5 decision, Mohd Salim, through his lawyer MC Pant, approached the court again. While the lawyer did not specifically ask for legal personhood for the Ganga, he directed the court's attention to the case of the Whanganui River in Aotearoa New Zealand, which was given legal personhood through the Te Awa Tupua Act, approved by Parliament on March 14, 2017 (Jolly and Menon 2021, 469, Nakazora 2023, 181–182). In response to this approach over non-implementation and expressing its displeasure with the government's lack of action, the Uttarakhand High Court made its own landmark decision on March 20, 2017, declaring the Ganga and Yamuna rivers to be living entities "with all corresponding rights, duties and liabilities of a living person". Also, this decision was penned by Justice Sharma. In the following, it is referred to as the "Rivers" decision, or case.

In the Lalit Miglani case, the directives of the December 2 judgment had not been implemented either. The court expressed its "extreme displeasure" over this in new orders given on March 6 and March 17, 2017 (Upadhyay and Nayak 2024). Subsequent to the decision of giving legal personhood to the Ganga and Yamuna, Lalit Miglani filed a miscellaneous petition to have the Himalayas, glaciers, streams, water bodies and other natural entities to be declared juristic persons on par with the rivers Ganga and Yamuna. And on March 30, 2017, the Uttarakhand High Court declared precisely this, with Justice Sharma again authoring the decision. Like the other decisions, it was made with Justice Alok Singh as the other presiding judge. In the following, this decision is referred to as the "Glaciers" decision.

The state government of Uttarakhand challenged the Rivers judgment on the grounds that jurisdictional issues and matters of liability made it difficult to implement. The Supreme Court responded by staying the verdict on July 7, 2017. The central government followed suit by challenging the Glaciers decision on October 26, 2017, and the Supreme Court stayed it also. Both judgments are thereby suspended, pending a final decision in the Supreme Court. When this will happen – if ever – is an open question. There do not seem to be expectations that the Supreme Court will open proceedings over this in the near future.

While the decisions have been stayed, they are not overturned. They still set legal precedent and remain binding for lower courts and tribunals in Uttarakhand. This will remain the case at least until the Supreme Court hears the challenges to the decisions and comes to a decision of its own (Jolly and Menon 2021, 474).

The rulings

Neither of these judgments can be considered as being among the most elegant of legal texts. In particular the Glaciers decision, which runs to 66 pages,[2] the bulk of which consists of texts copied from various (and varied) sources, often without accompanying comments from the judge and is sometimes hard to follow. In the following, the judgments are summarized and compared, with the aim of clarifying the arguments and intentions behind the pronouncements.

While both rulings declare natural entities to be legal persons, there are significant differences in the way in which this directive is justified. Still, this chapter reads them together. As the conclusions in both cases are both radical and almost identical, and as they were both decided within a very short period of time, and by the same judges, we think that rather than focusing on the differences in arguments between the cases it is better to see them as complementary. Taken together, they can sum up the judges' full justification for declaring the rivers and glaciers (and other elements of the environment) as legal persons. This way of reading is meant to help reveal the intentions of the author behind the ruling, who, in this case, is Justice Sharma, since both the Rivers and Glaciers orders were authored by him.

The Rivers case

In the Rivers ruling, the Uttarakhand High Court expresses its "serious displeasure" with the way the state governments of Uttarakhand and Uttar Pradesh have failed to follow up the orders from the court's decision of December 5, 2016 (n. 9). It goes on to state that

> The extraordinary situation has arisen since Rivers Ganga and Yamuna are loosing [sic] their very existence. This situation requires extraordinary measures to be taken to preserve and conserve Rivers Ganga and Yamuna.
>
> (n. 10)

> Rivers Ganges and Yamuna are worshipped by Hindus. These rivers are very sacred and revered. The Hindus have a deep spiritual connection with Rivers Ganges & Yamuna. According to Hindu beliefs, a dip in River Ganga can wash away all the sins. The Ganga is also called 'Ganga Maa'. …
>
> (n. 11)

This is followed by several pages describing legal precedents where idols and deities have been recognized as legal or juristic persons. The argument is made that "Recognition of an entity as juristic person is for subserving the needs and faith of society" (n. 14).

Figure 4.2. The Ganga being formed at the confluence of the rivers Bhagirathi and Alaknanda at Devprayag in Uttarakhand, India. Photo by Kavita Upadhyay.

> In other words, the entity acts like a natural person but only through a designated person, as their Lordships have held in the judgments cited hereinabove, that for a bigger thrust of socio-political-scientific development, evolution of a fictional personality to be a juristic person becomes inevitable. This may be any entity, living inanimate, objects or things. …
>
> Thus, to protect the recognition and the faith of society, Rivers Ganga and Yamuna are required to be declared as the legal persons/living persons. (n. 16)

> Rivers Ganga and Yamuna are central to the existence of half of Indian population and their health and well being. The rivers have provided both physical and spiritual sustenance to all of us from time immemorial. Rivers Ganga and Yamuna have spiritual and physical sustenance. They support and assist both the life and natural resources and health and well-being of the entire community. Rivers Ganga and Yamuna are breathing, living and sustaining the communities from mountains to sea. (n. 17)

> Accordingly, while exercising the parens patrie jurisdiction, the Rivers Ganga and Yamuna, all their tributaries, streams, every natural water flowing with flow continuously or intermittently of these rivers, are

> declared as juristic/legal persons/living entities having the status of a legal person with all corresponding rights, duties and liabilities of a living person in order to preserve and conserve river Ganga and Yamuna. The Director NAMAMI Gange, the Chief Secretary of the State of Uttarakhand and the Advocate General of the State of Uttarakhand are hereby declared *persons in loco parentis* as the human face to protect, conserve and preserve Rivers Ganga and Yamuna and their tributaries. These Officers are bound to uphold the status of Rivers Ganges and Yamuna and also to promote the health and well being of these rivers.
>
> (n. 19)

The Glaciers case

The Glaciers ruling starts by arguing that in PILs the court has considerable leeway to rule on cases that have societal significance and to depart from standard legal procedures (p. 4). It then goes on to state the seriousness of the condition of the glaciers Gangotri and Yamunotri, sources of the rivers Ganga and Yamuna, respectively, which are receding at alarming rates. Also, it is stated that the "Ganga and Yamuna are revered as deities by Hindus" (p. 6).

From page 6 to page 59, the document consists almost exclusively of texts culled from elsewhere. First are a number of extracts from articles on the social and spiritual value of nature (trees in particular) by writers and activists from different parts of the world. Then there are a number of pages excerpted from a natural science text on biodiversity by "learned authors". This is followed by extensive sections from international conventions and declarations (The Stockholm Declaration from 1972, the UN World Charter for Nature (1982), The Rio Declaration (1992), The Convention on International Trade in Endangered Species (1973), the Rio Forest Principles (also from Rio 1992)). There are passages from texts on the Chipko movement, including songs translated into English and the Tilari declaration of the people of Tehri-Garhwal in 1968. There is a very brief reference to the Te Urewera case in Aotearoa New Zealand (declaring a forest a legal person in 2014). On page 42, there is a paragraph apparently in the court's own words, which states:

> It is the fundamental duty of all the citizens to preserve and conserve the nature in its pristine glory. There is a grave threat to the very existence of Glaciers, Air, Rivers, rivulets, streams, Water Bodies including Meadows and Dales. The Court can take judicial notice of the fact that few cities are not liveable due to higher level of pollutants in the atmosphere. The Courts are duty bound to protect the environmental ecology under the 'New Environment Justice Jurisprudence' and also under the principles of parens patriae.

Thereafter follows close to 18 pages of a legal discussion of the parens patriae principle in the US, with extensive references to US case law. The source of this text is not given. A paragraph from a text on the relationship between morals and law follows.

Pages 61 and 62 conclude the arguments of the court. Key parts of this text are reproduced here:

> Rivers and Lakes have intrinsic right not to be polluted. Polluting and damaging the rivers, forests, lakes, water bodies, air and glaciers will be legally equivalent to harming, hurting and causing injury to person. Rivers, Forests, Lakes, Water Bodies, Air, Glaciers and Springs have a right to exist, persist, maintain, sustain and regenerate their own vital ecology system. The rivers are not just water bodies. These are scientifically and biologically living. The rivers, forests, lakes, water bodies, air, glaciers, human life are unified and are indivisible whole. The integrity of the rivers is required to be maintained from Glaciers to Ocean.
>
> We must recognize and bestow the Constitutional legal rights to the 'Mother Earth'.
>
> The very existence of the rivers, forests, lakes, water bodies, air and glaciers is at stake due to global warming, climate change and pollution. Trees are the buffer zone necessary to protect the glaciers from direct and indirect heat. One tree sustains life of thousand[s] of insects. Birds chirp and make their nests on the trees. Trees are mini-reservoirs and have a capacity to store the water. The water stored by the trees is released slowly. The Oak tree preserves about 75,000 gallon[s] of pure water. Plucking of one leaf, grass blade also damages the environment universally.
>
> The leading civilizations have vanished due to severe droughts. Water is elixir of life and we must conserve and preserve every drop of water. The value of water should not be undermined only for the reason that it is still available in plenty. The past generations have handed over the 'Mother Earth' to us in its pristine glory and we are morally bound to hand over the same Mother Earth to the next generation.

From this, a series of directions are issued. Legal personhood is accorded in this way:

> We, by invoking our parens patriae jurisdiction, declare the Glaciers including Gangotri & Yamunotri, rivers, streams, rivulets, lakes, air, meadows, dales, jungles, forests wetlands, grasslands, springs and waterfalls, legal entity/legal person/juristic person/juridical person/moral person/artificial person having the status of a legal person, with all corresponding rights, duties and liabilities of a living person, in order to preserve and conserve them. They are also accorded the rights akin to fundamental rights/legal rights.

Furthermore, persons in loco parentis or the human face, tasked with protecting the glaciers and other natural elements, are named who are the same three persons as in the Rivers case, plus others, including the Director (Academics) at the Chandigarh Judicial Academy, as well as veteran environmental lawyer MC Mehta. Moreover, "The Chief Secretary of the State of Uttarakhand is also permitted to co-opt as many as Seven public representatives from all the cities, towns and villages of the State of Uttarakhand to give representation to the communities living on the banks of rivers near lakes and glaciers".

To sum up

The court decisions are premised on a number of postulates:

- Rivers and glaciers being of great importance for human life.
- Particular importance of the rivers as sacred to Hindus.
- Interdependence of ecosystems and all their elements, including humans.
- Moral obligation to protect (Mother) nature also for its own sake (which implies the inherent value of nature and, from a legal perspective, the rights of nature).
- Existential threats to the rivers, glaciers and connected ecosystems.

These factors are seen as obliging the court to act, with reference to the parens patriae doctrine: the role of the state as protector of those who cannot protect themselves. The decisions employ the legal fiction of designating the rivers, glaciers and ecosystems as legal persons and accord them all the rights of a living person. To represent these legal persons, public officials are designated to act *in loco parentis*, as guardians. (In the Glaciers case, the public officials are complemented with community representatives.)

Background for the decisions

Declaring natural entities – rivers, glaciers etc. – to be legal persons, with all the rights, duties and liabilities of a human being, as the Uttarakhand High Court did, is a radical break with Indian environmental jurisprudence (Jolly and Menon 2021, 471).

These decisions do not follow logically as responses to the original petitions and grievances of the cases. Mohd Salim wished to have constructions removed and informal settlers evicted along the river and did not ask to have the two rivers of more than 3,500 km in total length declared to be legal persons with corresponding rights. Lalit Miglani initially wished to remedy the contamination of the Ganga rather than address the situation of the glaciers from which the Ganga as well as the Yamuna originate.

Moreover, Upadhyay and Nayak conclude that the relevance of the decisions for the specific challenges of Uttarakhand is limited. While the Rivers decision does address the issue of contamination of the Ganga, which is a concern in the lower reaches of the river, "the rulings fail to mention pressing issues such as floods, landslides, infrastructure projects like dams and roads, climate change-related disasters and their interconnectedness" (Upadhyay and Nayak 2024, 15).

Ideas of legal personhood and rights of nature had not been widely discussed and disseminated in India before the Rivers and Glaciers decisions. An initiative for a Ganga Rights Act had been launched by the Ganga Action Parivar – an alliance of faith-based organizations and individuals – in 2016, but it did not achieve much traction or publicity.[3] There are no indications that the court was aware of this initiative or was influenced by it. More visible initiatives for protecting and cleaning the Ganga – such as the campaigns and hunger strikes by Swami Nigamananda and by GD Agarwal mentioned above – did not lobby for legal personhood for the river. Agarwal's activism, for instance, was consistently referring to the religious importance of the river and the Hindu duty to protect it. Even though he was aware of the Rivers case (Ranjan 2024), Agarwal did not refer to it or ask for legal rights for the river during his last hunger strike, which ended with his death in 2018. While environmental activists in India were generally happy with the intention behind the rulings, scepticism was expressed about whether they would have any positive impact for the Rivers (Goswami 2017, Kothari and Bajpai 2017, Dutta 2022).

Given the fact that the decisions have been challenged in the Supreme Court by both the Uttarakhand state government and by the central government, it does not seem that the Justices Sharma and Singh made their pronouncements on the basis of an interest from the side of the state either. Thus, put bluntly, the radical decisions of giving personhood to nature were not asked for in the original petitions, nor did they connect logically with the original claims of these cases or seem particularly relevant for the environmental concerns of the state, and they were made without any firm grounding in civil society activism or support from the state and its bureaucracy.

India has a strong tradition of judicial activism (Maté 2015, 171). PILs have been an arena for judicial activism, allowing judges to give injunctions to state institutions for redressing violations of fundamental rights and to interpret and extend existing law. The Indian PIL has been praised as an example that other countries ought to emulate (Holladay 2012), while for other legal scholars, the model implies a disregard for democratic legislative process and "has provided judges with unchecked power which they deploy in capricious ways" (Chandra 2018, 712). Relaxed requirements of standing allow any public-spirited person to file a petition on behalf of those who cannot, and the liberal procedures also permit judges extensive authority to pursue matters *suo motu* – that is, on their own initiative, without a specific petition being filed (Alley 2019, 6). There is a long tradition of activist judges pushing the law to expand its scope and reach. This is important context for understanding these Uttarakhand High Court rulings. In the opening statements of the Glaciers order, the judges argue that

> in a public interest litigation, unlike traditional dispute resolution mechanism, there is no determination of adjudication of individual rights. …
>
> More importantly, the court is not merely a passive, disinterested umpire or onlooker, but has a more dynamic and positive role with the responsibility for the organisation of the proceedings, moulding of the relief and – this is important – also supervising the implementation thereof.
>
> Their Lordships of Hon. Supreme Court in (1989) Supp (1) SCC 504 in the case of 'Rural Litigation & Entitlement Kendra v. State of U.P.' have held that in matters of grave public importance, *court is not bound by procedural technicalities.*
>
> (Lalit Miglani v. State of Uttarakhand & Others 2017, p. 4, emphasis added)

Within this judicial tradition, and with the understanding that the rivers Ganga and Yamuna were in a critical situation, the court saw this as an instance where it was obliged to act on behalf of rivers, which are unable to approach the courts themselves, and where it could and should go beyond the scope of the original PIL petitions.

Many have pointed to the personal role of Justice Sharma for the rulings. After the Rivers and Glaciers cases, he granted legal personality to nature in several other cases. While still at Uttarakhand High Court, he decided a case about the treatment of horses and declared that

> The entire animal kingdom including avian and aquatic are declared as legal entities having a distinct persona with corresponding rights, duties and liabilities of a living person. All the citizens throughout the State of Uttarakhand are hereby declared persons in loco parentis as the human face for the welfare/protection of animals.
>
> (Narayan Dutt Bhatt v. Union of India 2018)

After he was transferred to the Punjab and Haryana High Court in 2018, he repeated this Animal ruling also for this new jurisdiction. And in 2020, he granted legal personhood to another water body, Lake Sukhna. These are by no means the only rulings of Justice Sharma that stand out. Several newspaper and magazine reports detail tens of cases where he has made surprising and far-reaching decisions (Vishwanath 2018, Sethi 2020, The Tribune 2020). He has demonstrated willingness to rule not only on environmental issues and animal rights but has also addressed drug problems, noise pollution, workers' welfare provisions, caste discrimination, and right to education and health, including mental health, and lots of other issues. Not without reason he has been called "a crusader judge" and billed as "India's new 'green judge'" (Nakazora 2023, 182).

His swiftness in producing his rulings has been commented upon: "Matters pending for decades were decided after hearings lasting maximum up to three hours" (The Tribune 2020). This speed in deciding cases has been qualified as haste that impacts the quality of his decisions, with reference to the Rivers and Glaciers cases (Reddy 2018, 6). Vishwanath quotes lawyers saying that the many decisions of Justice Sharma that go beyond the scope of the original cases are not an aberration within India's courts, but rather the outcome of a system that produces "attention-grabbing judges". An advocate at the Uttarakhand High Court who declined to be identified is quoted as saying:

> This is a small high court and not more than one judge makes it to the Supreme Court at any point of time, so the judges remain here till retirement. Something has to be done to stand out and be noticed. There are at least 3–4 environmental PILs in this court listed almost every day, and each case is an opportunity.
>
> (Vishwanath 2018)

This is not meant to sow doubts about the motives of Judge Sharma or imply that he was simply after being appointed to the Supreme Court. Unfortunately, he declined to be interviewed for this study, so we have not had any direct contact with him. But from statements from his colleagues and newspaper reports (Vishwanath 2018, Sethi 2020, The Tribune 2020), it seems beyond doubt that his concern for Ganga and for the environment is real. Rather, the preceding paragraph is meant to point to the systemic issue of Indian courts that has helped create the particular procedures and legal attitudes that facilitate judicial activism.

Another factor that undoubtedly has contributed to the Rivers and Glaciers orders is the court having been made aware of the legislation being approved in Aotearoa New Zealand, especially the case of the Whanganui River. Several studies confirm that the lawyer of Mohd Salim – Pant – made the judges aware of the Whanganui case before the March 2017 decision (Alley 2019, Nakazora 2023). The Whanganui case is not mentioned in any of the rulings; there is only a brief reference in the Glaciers decision to the Te Urewera Forest in Aotearoa New Zealand having been given legal personhood. But in the Rivers decision, traces of the language from the Te Awa Tupua (Whanganui Settlement) Act can be found. In the directives, the Ganga and Yamuna are spoken of as giving all of us "physical and spiritual sustenance", the exact same phrasing that is found in the Te Awa Tupua Act. The people to act as guardians are termed "the human face" of the rivers, which is the same expression that is used in the Whanganui Act. And the rivers are spoken of as wholes "from mountain to sea", again repeating one of the central expressions (or *kawa*) for Whanganui. Awareness of the Aotearoa New Zealand case is likely to have influenced these decisions.

The rulings by Justices Sharma and Singh cannot be explained simply as outcomes of the context in which they were made. They must rather be seen as radical breaks with existing jurisprudence, and which have only tenuous connections to the petitions that they address. But the Indian tradition of judicial activism and leeway for judges to go beyond the scope of petitions in

PILs as well as the awareness of the Whanganui and Te Urewera cases of rights for nature legislation in Aotearoa New Zealand are factors that facilitated the decisions of the Uttarakhand High Court.

Questioning the Rivers and Glaciers rulings

The case of Ganga and Yamuna personhood differs in many ways from the other cases of river persons we discuss in this book, including by the fact that the rulings have been stayed and are therefore not being implemented. We have consequently not been able to investigate empirically the impacts of these decisions. Gauging their possible strengths and weaknesses therefore becomes a more theoretical exercise. For this it is useful to review the way the rulings have been received and commented upon, by authorities and by academics.

Jurisdiction

The Rivers ruling was challenged in a Special Leave Petition to the Supreme Court by the Uttarakhand state government. The grounds for appealing the decision referred to legal hurdles arising from the fact that the rivers flow through several Indian states, and the inability of Uttarakhand officials – who are named as the human face of the rivers – to enforce the judgment in other states (State of Uttarakhand v. Mohd. Salim 2017). It was argued that the Uttarakhand High Court had exceeded its jurisdiction, also since the Constitution of India mandates it as the sole right of the Union Government to develop rules for the management of inter-state rivers. The appeal also raised the question of the liability of the guardians, and whether they, for instance, could be sued for damages in the case of human casualties in a flood (Singh 2017).

The question of jurisdiction becomes even trickier when considering that the Ganga also crosses the border into Bangladesh before eventually ending in the Bay of Bengal, and that the ruling talks about the river "from mountain to sea".

Guardians

These issues are closely linked to other concerns that have been raised over the "human face" arrangements mandated by the Uttarakhand High Court. For instance: if the responsible officials of the Uttarakhand state government have failed to protect the river as the existing legislation requires, what sense does it make to name these same persons as guardians of a new order that is probably even more challenging to uphold (Chicu 2017, Kothari and Bajpai 2017, O'Donnell 2018, Kinkaid 2019, 561)? Indeed, when the state government is both a violator of the rights of the rivers and its designated protector, there is a clear conflict of interest (Blair 2023, 7). A related but different concern is also raised by Jolly and Menon. They point out that designating state officials as the human face of the rivers bestows a superior *locus standi* on these representatives and may result in tightened standing requirements before the courts. They argue that it may become impossible for concerned citizens to file PIL petitions on behalf of the rivers when there are public officials who have been mandated to act *in loco parentis* to protect, conserve and preserve the rivers (Jolly and Menon 2021, 484). If these officials fail to exercise their role as protectors, Ganga's protection in courts may in practice end up being weakened by the guardian arrangement.

In the Glaciers decision, the composition of guardians is expanded somewhat. The number of named guardians is increased from three to seven, and one of the new ones is not a public official but the veteran environmental lawyer and PIL petitioner MC Mehta. Moreover, it is argued that "the local inhabitants living on the banks of rivers, lakes and whose lives are linked with rivers and lakes must have their voice too" (Lalit Miglani v. State of Uttarakhand & Others 2017, p. 61), and the directives add:

> The Chief Secretary of the State of Uttarakhand is also permitted to co-opt as many as Seven public representatives from all the cities, towns and villages of the State of Uttarakhand to give representation to the communities living on the banks of rivers near lakes and glaciers.

The intention seems to be that these seven society representatives are also to act as the "human face" of the glaciers and other natural entities, but this is not stated explicitly. No description of the roles of these seven citizens is given, nor of how they are to be selected. Thus, the Glaciers decision cannot be said to evade the shortcomings pointed out above, even if the guardian arrangement is slightly expanded from the previous Rivers decision.

There is another argument about the relevance of legal personhood for nature in the Indian context that also involves the role of the state. It states that a radical break with existing jurisprudence – such as giving legal personhood to nature – should only be considered if it can be shown to improve outcomes in comparison to existing legislation. According to Jolly and Menon, the legal doctrines underlying India's environmental legislation – the polluter pays principle, and the public trust doctrine (that the state holds the environment in trust for the people) – give the necessary instruments for protecting nature if they are followed (Jolly and Menon 2021, 489–491). Thus, if existing legislation is not implemented, why should one expect better environmental outcomes by introducing new legal principles (that may be even more difficult to uphold)?

In several ways, the rulings are quite vague as to how they are supposed to be implemented and the effects they are supposed to have. As mentioned, they offer no guidance on how the guardians are to operate or achieve their mandate of protection and preservation. There is also no allocated budget for the costs involved. Another side to this is that they are vague in terms of what rights are accorded to the rivers and glaciers, and even less so in terms of duties and responsibilities.

Rights and duties

O'Donnell points to the difficulties that result from giving rivers the same rights as living persons. According to the Glaciers ruling: "Polluting and damaging the rivers, forests, lakes, water bodies, air and glaciers will be legally equivalent to harming, hurting and causing injury to person. … Plucking of one leaf, grass blade also damages the environment universally". Taken literally, the ruling is impossible to implement. O'Donnell criticizes the rulings for conflating the legal person and the living person, thereby also failing to distinguish nature's rights from human rights when giving rights to rivers (O'Donnell 2018, 138). A legal person can be given specific rights tailored to the kind of entity it is given to. This was suggested by Christopher Stone already in 1972, when he even foresaw that different types of natural entities could be accorded different types of rights (Stone 1972, Jolly and Menon 2021, 482; see also O'Donnell 2018, 138–139).

The rivers Ganga and Yamuna, and nature, are also assigned the "duties and liabilities of a living person". It is difficult to understand what legal duties the court thinks natural entities can have. Moreover, liabilities could imply that individuals named as guardians would be responsible for damage or drowning from flood, as questioned by the Uttarakhand state government in their appeals. Jolly and Menon claim that many of the objections raised here come from the court's lack of engagement with the international experiences and discussions on rights of nature jurisprudence (Jolly and Menon 2021, 487).

Religion

Ideas surrounding purity and pollution are complex and crucial in India and are dealt with elsewhere in this book (see Brandshaug et al. this volume, Borchgrevink this volume). Here we will briefly discuss what has been pointed out as problematic by several writers: the underpinning of the decisions with reference to Hindu values and beliefs (i.e., Kothari and Bajpai 2017, O'Donnell 2018, Kinkaid 2019). This contradicts the secular nature of the Indian Constitution and jurisprudence, in that it fails to acknowledge the many non-Hindus living within the river basin and therefore may potentially exclude them from rights to the Ganga. In the context of the present Hindu nationalist government of the Bharatiya Janata Party (BJP) and Prime Minister Narendra Modi, such concerns over marginalization of Muslims and others take on greater significance (Borchgrevink this volume).

However, reading the rulings closely – in particular the Glaciers ruling – to some extent exonerates them from charges of being expressions of political Hindu nationalism. It is clear that Hinduism is one among several inspirational strands that the decisions rest on. The sacredness of the Ganga and Yamuna is mentioned in both decisions, although only briefly. However, in the Glaciers decision, nine tenths of the pages are dedicated to presenting a broader environmentalism as justification for giving legal personhood to nature. From the many sources the judges have selected to quote in this ruling, we can glean an ideology that rests on the idea of the fundamental importance of the natural environment for human life and the interconnectedness of all of nature. It is certainly a spiritual form of environmentalism, which argues for the intrinsic value of nature and our duty to protect it, but it is not tied to only one particular religion. Alley (2019) and Nakazora (2023) both argue against reading these pronouncements as expressions of *Hindutva*, or Hindu nationalism (although Alley does speak of "Hindutva lite"). Nakazora points out that Mohd Salim, the petitioner whose lawyer suggested legal personhood for nature, is a Muslim and was happy with the verdict, as further evidence of the case having other roots than Hinduism (Nakazora 2023, 182).

Even if Hindu nationalism was not the rationale for the court's granting of legal personhood to the Ganga and Yamuna, they might still be used for this purpose. Given the importance of the rivers for Hindus, the Ganga in particular, the personhood and legal rights for the rivers could easily be used to promote the ideology. Indeed, in Modi's early campaigns, cleaning the Ganga was among the issues that was high on the agenda. Still, in practice, the rulings have not been made use of by Hindutva organizations, at least not so far. Neither the BJP nor its associated organizations have embraced the judgments. In fact, the opposite has taken place. The appeals to the Supreme Court to have the decisions overturned came from the BJP governments at state and central levels (Nakazora 2023, 179).

Human vs nature's rights

A concern sometimes raised over giving rights to nature is that they may come into conflict with human rights, in particular for vulnerable groups such as Indigenous peoples and others (O'Donnell and Macpherson 2018, Marshall 2019). One can easily imagine such cases arising from the Rivers and Glaciers rulings that would involve different types of marginalized groups. The injunction to remove illegal settlements along the river is directed at migrants with minimal resources. Also, by foregrounding the sacredness of the rivers to Hindus in the rulings, the rights of non-Hindus to use the river might be limited. There may be precedents for this, as it has been argued that earlier measures to clean the Ganga through the closing down of polluting tanneries along the river have disproportionally targeted Muslims and Dalits, showing ways in which both religious minorities and marginalized groups, including lower-caste Hindus, have been impacted (Kinkaid 2019, 565). And the Glaciers decision singles out beggars as those to be expelled, by the injunction that "The District Magistrate, Haridwar is directed to ensure that the Beggars are not allowed to be present on the Ghats" (Lalit Miglani v. State of Uttarakhand & Others 2017, p. 64). (Ghats are areas along rivers from where rivers can be accessed. They may include steps leading down to the river.) Thus, there are real concerns that the rulings, should the stay be lifted, could adversely affect the rights of several marginalized groups.

Conclusion

How should we assess the Uttarakhand rulings in relation to the global ideas of nature's rights? One observation is that the Glaciers decision combines features from the two ideal models for rights of nature described by Kauffman and Martin (2021; see also Brandshaug and Borchgrevink this volume). They distinguish between the "Personhood Model" that gives the rights of a legal person to a particular natural entity, and the broader "Nature's Rights Model" that gives a different type of rights – "existence rights" – to all of nature. In the ruling, the two models are mixed by granting personhood rights to "the Glaciers including Gangotri & Yamunotri, rivers, streams, rivulets, lakes, air, meadows, dales, jungles, forests wetlands, grasslands, springs and waterfalls". This may not comprise all of nature, but it comes close and is certainly not easily delimited as a natural entity. Personhood is awarded to something that resembles what Tanasescu identifies as "Nature with a capital N" (2022, 32; see also Brandshaug and Borchgrevink this volume). This is nature as a totality, abstracted from the particular.

The judgment presents an understanding of nature as an equilibrium system that we are duty-bound to preserve, one where all elements are balanced and in dependence of each other.

> The rivers, forests, lakes, water bodies, air, glaciers, human life are unified and are [an] indivisible whole … Plucking of one leaf, grass blade also damages the environment universally … It is the fundamental duty of all the citizens to preserve and conserve the nature in its pristine glory … The Courts are duty bound to protect the environmental ecology under the 'New Environment Justice Jurisprudence'.
> (Lalit Miglani v. State of Uttarakhand & Others 2017, pp. 61–62, 42)

Here, the court explicitly places itself in the eco-theological tradition (Tanasescu 2022) of Thomas Berry (2009) and Cormac Cullinan (2011).

This understanding of nature as a finely balanced system where even the plucking of one leaf damages the environment universally implies an extremely broad notion of harm that leaves little room for human livelihoods. Moreover, all harm to nature should be prosecuted:

> Any person causing any injury and harm, intentionally or unintentionally to the Himalayas, Glaciers, rivers, streams, rivulets, lakes, air, meadows, dales, jungles and forests is liable to be proceeded against under the common law, penal laws, environmental laws and other statutory enactments governing the field.
> (Lalit Miglani v. State of Uttarakhand & Others 2017, p. 63)

As O'Donnell points out, the extremely broad notion of harm, and the corresponding injunction to prosecute all harm, makes it difficult to envision successful implementation of these decisions even if they had not been stayed (O'Donnell 2018).

To conclude, the Rivers and Glaciers rulings are landmark cases by virtue of having opened Indian legislation to the possibility of giving rights and personhood to nature. But there are serious weaknesses in the institutional arrangements, specification of rights and underlying justifications of the Uttarakhand High Court decisions. When – or if – the Supreme Court deliberates on the appeals, they may find ways of overcoming these weaknesses.

Until that happens, the rulings are stayed and have no direct effects. But there have been indirect effects. They have inspired further attempts at giving nature rights in India, through new court decisions. Some of them – as referred to above – have been made by the same judge, Justice Sharma, but not all. In April 2022, the Madras High Court declared Mother Nature a living being. This decision was made by Justice S. Srimathy, and in her ruling, she referred to the Lalit Miglani case on which the Glaciers order was given (Periyakaruppan v. Revenue Department, Government of Tamil Nadu 2022). Moreover, in 2017, the legislative assembly of Madhya Pradesh passed the resolution granting legal personality to the Narmada River (Madhya Pradesh Vidhan Sabha 2017). This resolution has been seen as inspired by the Rivers decision of the Uttarakhand High Court (Blair 2023).

Notes

1 For instance: Maneka Gandhi v Union of India AIR 1978 SC 597; Kharak Singh v State of Uttar Pradesh AIR 1963 SC 1295; Sunil Batra v Delhi Administration AIR 1978 SC 1675; Francis Coralie Mullin v Administrator, Union Territory of Delhi AIR 1981 SC 746; Unni Krishnan v State of AP AIR 1993 SC 645; People's Union for Civil Liberties v UOI AIR 2001 SC 454; Narmada Bachao Andolan v Union of India AIR 2000 SC 375; MC Mehta v Union of India (2004) 3 SCR 128.

2 Downloading the ruling from the webpage can yield a different number of pages for the document. The page numbers in the following refer to the version with a total of 66 pages.

3 Although Kothari and Bajpai claim the national government were considering approving the initiative to give legal rights to the Ganga (2017).

References

Alley, Kelly D. 2019. "River goddesses, personhood and rights of nature: Implications for spiritual ecology." *Religions* 10 (9): 502.

Berry, Thomas. 2009. *The sacred universe: Earth, spirituality, and religion in the twenty-first century*. Columbia University Press.

Blair, Harry. 2023. "Saving India's rivers: Ecology, civil society, religion, and legal personhood." *World Development Sustainability* 3, 100068.

Chandra, Aparna. 2018. "Courting the people: Public Interest Litigation in post-emergency India." *International Journal of Constitutional Law* 16 (2): 710–718, doi:10.1093/icon/moy045.

Chicu, Lokgariwar. 2017. "The sad state of these persons called Ganga & Yamuna – Can state protect them?" *South Asian Network on Dams, Rivers and People (SANDRP)*. April 11. https://sandrp.in/2017/04/11/the-sad-state-of-these-persons-called-ganga-yamuna-can-state-protect-them/#more-22694.

Cullinan, Cormac. 2011. *Wild law: A manifesto for earth justice*. Bloomsbury Publishing.

Dutta, Ritwick. 2022. "Rights for 'Mother Nature' only feeds the myth of sustainable development." *The Wire*, May 8. https://science.thewire.in/environment/mother-nature-rights-madras-high-court-parens-patriae/.

Goswami, Urma A. 2017. "Will granting legal rights to rivers like the Ganga, change the on-ground situation?" *The Economic Times*, March 25. https://economictimes.indiatimes.com/news/politics-and-nation/will-granting-legal-rights-to-rivers-like-the-ganga-change-on-ground-situation/articleshow/57818653.cms?from=mdr.

Holladay, Zachary. 2012. "Public interest litigation in India as a paradigm for developing nations." *Indiana Journal of Global Legal Studies* 19 (2): 555–573.

IPCC. 2021. *The physical science basis: Contribution of Working Group I to the Sixth Assessment Report of the Intergovernmental Panel on Climate Change*. https://www.ipcc.ch/report/sixth-assessment-report- working-group-i/.

Jolly, Stellina, and K.S. Roshan Menon. 2021. "Of ebbs and flows: Understanding the legal consequences of granting personhood to natural entities in India." *Transnational Environmental Law* 10 (3): 467–492.

Kauffman, Craig M., and Pamela L. Martin. 2021. *The politics of rights of nature: Strategies for building a more sustainable future*. MIT Press.

Kinkaid, Eden. 2019. "'Rights of nature' in translation: Assemblage geographies, boundary objects, and translocal social movements." *Transactions of the Institute of British Geographers* 44 (3): 555–570.

Kothari, Ashish, and Shrishtee Bajpai. 2017. "We are the river, the river is us." *Economic & Political Weekly* 52 (37): 103–109.

Lalit Miglani v. State of Uttarakhand & Others. 2015. August 19, for Writ Petition (PIL) No. 140 of 2015.

Lalit Miglani v. State of Uttarakhand & Others. 2016. December 2, for Writ Petition (PIL) No. 140 of 2015.

Lalit Miglani v. State of Uttarakhand & Others. 2017. March 30, for Writ Petition (PIL) No. 140 of 2015.

Macpherson, Elizabeth, Axel Borchgrevink, Rahul Ranjan and Catalina Vallejo Piedrahíta. 2021. "Where ordinary laws fall short: 'riverine rights' and constitutionalism." *Griffith Law Review* 30 (3): 438–473. doi:10.1080/10383441.2021.1982119.

Madhya Pradesh Vidhan Sabha. 2017. "Maa Narmada nadi ko jeevit ikaai ke roop mein vaidhaanik adhikaaron ke sanrakshan vidheyak." Madhya Pradesh Vidhan Sabha proceedings, May 3, 2017.

Marshall, Virginia. 2019. "Removing the veil from the 'Rights of Nature': The dichotomy between First Nations customary rights and environmental legal personhood." *Australian Feminist Law Journal* 45 (2): 233–248.

Maté, Manoj. 2015. "The rise of judicial governance in the Supreme Court of India." *Boston University International Law Journal* 33 (169).

Mohd. Salim v. State of Uttarakhand & Others. 2014. July 30, for Writ Petition (PIL) No. 126 of 2014.

Mohd. Salim v. State of Uttarakhand & Others. 2016. December 5, for Writ Petition (PIL) No. 126 of 2014.

Mohd. Salim v. State of Uttarakhand & Others. 2017. March 20, for Writ Petition (PIL) No. 126 of 2014.

Nakazora, Moe. 2023. "Environmental law with non-human features in India: Giving legal personhood to the Ganges." *South Asia Research* 43 (2): 172–191.

Narayan Dutt Bhatt v. Union of India. 2018. SCC Online Utt. 645., n. 99.

O'Donnell, Erin L. 2018. "At the intersection of the sacred and the legal: Rights for nature in Uttarakhand, India." *Journal of Environmental Law* 30: 135–144.

O'Donnell, Erin, and Elizabeth Macpherson. 2018. "Voice, power and legitimacy: The role of the legal person in river management in New Zealand, Chile and Australia." *Australasian Journal of Water Resources* 23 (1): 35–44.

Periyakaruppan v. Revenue Department, Government of Tamil Nadu. 2022. April 19, for W.P. (MD) Nos. 18636 of 2013 and 3070 of 2020.

Ranjan, Rahul. 2024. "Sadhu activism: Rights of rivers and facets of religious environmentalism in the Himalayas." *The International Journal of Human Rights*: 1–21. doi:10.1080/13642987.2024.2420244.

Reddy, Chandan B.R. 2018. "Legal personality to rivers: An in-depth analysis." *Pen Acclaims* 3: 1–15.

Sethi, Chitleen K. 2020. "HC judge who declared Sukhna lake living entity also granted legal rights to Ganga, animals." *The Print*, March 5. https://theprint.in/judiciary/hc-judge-who-declared-sukhna-lake-living-entity-also-granted-legal-rights-to-ganga-animals/375776/.

Singh, Soibam Rocky. 2017. "Uttarakhand govt moves SC against Ganga's living entity status." *Hindustan Times*, June 25. https://www.hindustantimes.com/india-news/uttarakhand-govt-moves-sc-against-ganga-s-living-entity-status/story-9QXgAu5nACW1IKpEEN5XxO.html.

State of Uttarakhand v. Mohd. Salim. 2017. June 22, for SLP (C) No. 016879 of 2017.

Stone, Christopher. 1972. "Should trees have standing?—Toward legal rights for natural objects." *Southern California Law Review* 45: 450–501.

Tanasescu, Mihnea. 2022. *Understanding the rights of nature: A critical introduction.* transcript Verlag.

The Tribune. 2020. "Judge who gave living entity status to Ganga, Sukhna Lake retires." October 8. https://www.tribuneindia.com/news/punjab/judge-who-gave-living-entity-status-to-ganga-sukhna-lake-retires-152490/.

Tripathi, Purnima S. 2011. "Unsung baba." *Frontline*, July 15. https://frontline.thehindu.com/other/article30176114.ece.

Upadhyay, Kavita. 2018. "Activist GD Agrawal on fast-unto-death for Ganga conservation dies of cardiac arrest." *The Indian Express*, October 12. https://indianexpress.com/article/india/activist-gd-agarwal-who-was-on-indefinite-fast-to-save-ganga-dies-of-heat-attack-5397603/.

Upadhyay, Kavita. 2024. "As norms are ignored, Uttarakhand faces tunnelling disasters." *The Third Pole*, January 31. https://www.thethirdpole.net/en/climate/as-norms-are- ignored-uttarakhand-faces-tunnelling-disasters/.

Upadhyay, Kavita, and Bibhu Prasad Nayak. 2024. "Tracing the legal journey of petitions in the Uttarakhand High Court that became springboards for rights of rivers and nature in India." *The International Journal of Human Rights*: 1–20. doi:10.1080/13642987.2024.2380850.

Vishwanath, Apurva. 2018. "The animal kingdom can defend itself now, and it's all thanks to Uttarakhand's crusader judge." *The Print*, July 22. https://theprint.in/india/governance/the-animal-kingdom-can-defend-itself-now-and-its-all-thanks-to-uttarakhands-crusader-judge/84710/.

World Bank. 2015. "The National Ganga River Basin Project." https://www.worldbank.org/en/news/feature/2015/03/23/india-the-national-ganga-river-basin-project.

5 People-river inseparability

Malene K. Brandshaug, Miriama Cribb, Jenny Moreno Socha, Daniel Alfonso León and Arlinton Cuesta Mosquera

Introduction

The rivers Whanganui in Aotearoa New Zealand, Atrato in Colombia and Ganga and Yamuna in India are home to animal and plant species and to the human communities living along their banks. The rivers are also considered spirits, ancestors, divinities or other kinds of other-than-human beings. In the three cases, these rivers, communities and ways of life are extremely different, in terms of scale, volume, flow, topography, population diversity, river practices, environmental state and ecology. However, despite these differences, the lives of peoples and rivers are in each case intimately woven together. The rivers are understood as complex ecosystems of which humans are just a part and as larger more-than-human communities. Many people living along Whanganui, Atrato, Ganga and Yamuna engage in meaningful interactions with their river based on values of care, reciprocity and respect, but, as we highlight below, there are also troubled sides to these relationships. People and river embody each other's traits and problems, and no individuals relate to the river in the same way. In this chapter, we consider the everyday and spiritual engagement of people with their rivers, recognizing that this is also part of the backstory of the legal rulings for personhood rights for these rivers, passed in 2016 in Colombia and in 2017 in India and Aotearoa New Zealand.

We describe people-river relationships in Aotearoa New Zealand, Colombia and India, and along the rivers Whanganui, Atrato, Ganga and Yamuna, by drawing on ethnographic vignettes and detailed descriptions that are based on participant observation, interviews and conversations with people who are intimately related to these rivers. We also rely on some of our own personal, lived experiences with these rivers and aim for local voices to be heard. We acknowledge that written language, and especially the academic genre, is not sufficiently suited to convey the width, nuances and full character of the rivers or the people-river relationships. To make up for these limitations we also include photographs from these rivers, local stories about them, poetry and excerpts from novels. Through these means, we aim to convey the inseparability of people and river in each case.

DOI: 10.4324/9781003581208-5

In the first three sections of the chapter, we describe differences in size and diversity of rivers and human populations, detail how and whose people-river relationships are referred to in the legal documents for rights of these rivers and discuss how relationality is treated in rights of nature debates. We then proceed to describe some of the human-river relationships that are found along the Whanganui, the Atrato and the Ganga and Yamuna rivers, and lastly reflect on to what extent the legal riverine rights decisions safeguard this relationship.

Rivers and riverine populations

Whanganui River

The Whanganui River, found in the western part of the North Island of Aotearoa New Zealand, flows 290 km from its source Tongariro (mountain), located in the centre of the North Island. It first moves north to west, then makes a turn at Taumarunui township and continues south towards the city of Whanganui, where it flows into the Tasman Sea. A large part of the river is in the ecologically rich Whanganui National Park of native lowland forest, while the part running through Whanganui City is channelled and surrounded by built infrastructure. Being a relatively calm river with few rapids, it is possible to travel the length of the river by boat or canoe, and people frequently paddle the upper to middle reaches of the river. The river was the supply artery for communities along its banks and once provided plentiful food for its population. However, the river has been altered from hydropower production, erosion, runoff from farmland, sedimentation, development projects, flooding and industrial toxic water. Despite this, today fish are still caught, people swim in the river and the river remains a major drawcard for tourism.

Along the river, the are several villages (*marae*) populated by Whanganui *iwi* (Māori tribe) and their subtribes (*hapu*), as well as land and towns colonized by settlers and inhabited by their descendants, known as *Pākehā* (non-Māori), which now constitute the majority population in this area (Cribb this volume). The Māori are the Indigenous people of Aotearoa New Zealand, coming from Polynesia in the 13th century, while Whanganui *iwi* are Māori tribes and sub-tribes of the Whanganui River. *Pākehā*, descending from the settlers, largely from the colonial power United Kingdom, populated the Whanganui area in the mid-1800s. Today, there are approximately 70,000 people living in the area. In the three districts Whanganui River is part of (Whanganui, Ruapehu and Stratford Districts), Māori make up 30% of the population, which is higher than their 18% total of the Aotearoa New Zealand population (Stats NZ 2023). The rest of the population consist of *Pākehā* (non-Māori) and includes descendants of settlers as well as recent immigrants from Asia, the Pacific Islands and other parts of the world. Whanganui *iwi* have fought for the sovereignty, protection and recognition of their relationship with the river from as early as 1873, when the first river claim emerged in petitions to parliament. Through the Treaty of Waitangi settlement process, Te Awa Tupua (Whanganui River Claims Settlement) Act was passed in 2017, granting the river legal personhood (Cribb this volume). The Whanganui River holds immense spiritual and cultural significance for Whanganui *iwi*, who consider it a sacred ancestor and central to their identity, culture and livelihood. It is this relationship that informs a significant part of the Te Awa Tupua Act and is a focus of this chapter.

Atrato River

The Atrato River, formed by its tributaries in the western slopes of the Andean Cordillera Occidental, flows 750 km north from the mountains to the sea, through lowland tropical rainforest. On the way, it carries with it a vast amount of sediment area, which turns the colour of the water muddy brown as it passes the city of Quibdó. Towards the end, Atrato forms a great river delta, before it runs into the Gulf of Urabá and the Caribbean Sea. The river also carries with it the remains of decaying plants, trees and fauna and the waste of the towns and villages along its banks, mixed with chemical pollutants produced by industrial uses of the river, such as heavy metals from illegal gold mining (McNeish and Moreno Socha 2024).

Figure 5.1. River boats on the banks of Atrato River, at Quibdó. Photo by John Andrew McNeish.

Hence, while the river has amongst the highest levels of biodiversity in the world, it also carries with it multiple problems for both humans and nature. In addition to severe ecological threats, the area faces issues with poverty, ethnic marginalization, violence and competing armed actors. The riverbanks of the Atrato are the ancestral territories of Afro-Colombian and Indigenous communities. Approximately 500,000 people live in the Chocó Department, of which Afro-Colombians descendants make up 8%, Indigenous peoples 10% and the remaining 3% are mestizos. 683 Afro-Colombian communities are organized into 64 community councils, and there are 125 Indigenous reserves (Gobernación del Chocó 2020, Vallejo Piedrahíta, McNeish, Moreno Socha, Cuesta Mosquera and Alfonso León this volume). These communities depend on the river's rich waters for fishing, free time and other subsistence and social and cultural practices. The livelihoods of these communities are shaped by the ecosystem of the river but also by its past and present legal and illegal economic use, including mining operations, sand extraction, drug trade and armed conflicts. The associated encroachments on and contamination of the river and its surroundings have destroyed a fragile ecosystem on which local communities rely for multiple aspects of their lives. In reaction to the serious socio-environmental issues, local organizations filed a legal case with the Colombian Constitutional Court, which resulted in the Atrato River becoming a subject of rights (McNeish and Moreno Socha 2024, Vallejo Piedrahíta, McNeish, Moreno Socha, Cuesta Mosquera and Alfonso León this volume).

Ganga and Yamuna

The Ganga flows more than 2,500 km through diverse geographies, ecologies and regions – from Uttarakhand state in western Himalayas and across northern India, before crossing the border to Bangladesh, then finally running into the Bay of Bengal in the northeastern part of the Indian Ocean. Yamuna is the largest tributary river of Ganga with its 1,376 km, running mostly in parallel with Ganga's upper half. Both rivers originate in the Himalayan glaciers in Uttarakhand and merge in Prayagraj in Uttar Pradesh state.

Ganga is the earth's most populous river, and its basin is home to about 400 million Indians, and additional millions in Nepal, Bangladesh and China (de Micheaux 2019, 187) – about 630 million in the whole river basin. Given that nearly 8% of the world's population lives by Ganga and its tributaries (Eck 2012, chap. 4), there is a tremendous diversity in terms of, among other things, ethnicity, religion, caste and class. Both rivers are considered sacred to Hindus and are visited by large numbers of pilgrims.

The Ganga River basin is heavily dammed and partly canalized. It is subject to severe pollution from industrial and human sources (Eck 2012, chap. 4), and to encroachments caused by mining of riverbed material, irrigation infrastructure, hydropower projects and extraction of groundwater. These environmental problems are coupled with climate change-related issues (de Micheaux 2019, 187). The pollution of Ganga and Yamuna is a commonly known concern that is addressed to varying degrees and by different means by governmental and civil society actors. In 2014 and 2015, two public interest litigations were filed against river encroachments and contamination, which ultimately led a judge in Uttarakhand High Court to declare the rights of Ganga and Yamuna rivers in March 2017 (Borchgrevink and Upadhyay this volume).

People-river inseparability in the riverine rights decisions

From Whanganui's 290 km with a population of less than 100,000, to Atrato's 750 km and population of half a million, and again to Ganga's 2,500 km and population of hundreds of millions, we are dealing with rivers of very different scales. The geography of the rivers and the diversity of the riverine populations – together with the historical, political and legal circumstances described in previous chapters – have influenced whose relationship to the rivers is reflected in the legal personhood decision, and how this is achieved. The legal formulations of riverine rights passed in Aotearoa New Zealand (2017), Colombia (2016) and India (2017) made efforts to incorporate local and Indigenous lifeworlds and differently express people-river inseparability.

The legal process by Whanganui *iwi* started as early as 1873 through parliament petitions, followed by legal proceedings pursued through the Native Land Court, the Native Appellate Court, the Supreme Court, the Court of Appeal and a royal commission of inquiry. In 1990, Whanganui *iwi* filed the Whanganui River claim to the Waitangi Tribunal, which was established in 1975 to investigate Crown breaches of the Treaty of Waitangi, including prejudicial treatment and loss for Māori. It is through the Treaty of Waitangi settlement process that the Te Awa Tupua Act was passed in 2017. A central aim of the Act was to acknowledge and encompass Māori and more specifically Whanganui *iwi* worldviews and values. The Whanganui *iwi* is a relatively unified group of Indigenous Māori and have practices, stories, principles and laws, such as *Tupua te Kawa*, that can be recognized and included in legal decisions under the Te Awa Tupua Act.

In the Aotearoa New Zealand case, the Te Awa Tupua Act grants legal personhood to the Whanganui River and is thorough and detailed in its efforts to encompass Māori principles and values as accurately as possible. However, there have been challenges of translation, not only in terms of language but also in interpretation – in conveying the broader meaning of complex concepts like *tikanga* and *kawa* (Cribb this volume). Although the concept of personhood does not capture how Māori relate to the river as a living entity or ancestor, the inclusion of *Tupua te Kawa*, Whanganui *iwi* laws or principles, is closer in appreciating and incorporating what is described as the inseparable relationship between Whanganui *iwi* and Whanganui River. Especially the principle "*Ko au te awa, ko te awa ko au* – I am the river and the river is me" (see below) – conveys and reaffirms Māori Whanganui interrelatedness with the river within formal law. The legal document translates the concept of Te Awa Tupua as "the inseparability of people and the River", acknowledges their "special relationship with the Whanganui River" and expresses respect for "the intrinsic connection between the Iwi and Hapu of Whanganui and the Whanganui River reflected in the Whanganui *pepeha*, 'Ko au te awa, ko te awa ko au'" (Te Awa Tupua Act 2017, 69 Acknowledgements). These acknowledgements and the extensive inclusion of Māori principles were made possible by the inclusion of Whanganui *iwi* themselves in an extensive participatory process.

In the case of the Atrato River in Colombia, diverse local groups and communities with people of Indigenous, Afro-descendants and mestizo ethnicity live along the course of the river. Each relate to the river in a different way, and the river people here are much less unified than the Whanganui Māori. Although people of the Atrato have managed to mobilize and make demands for the river across groups and many share a relational way of understanding and being with the river, the larger heterogeneity in interests, ways of life and connections to the river made it more challenging to echo locally specific people-river relationships in the legal framework. Although differently understood and expressed, there is nonetheless a common perspective of the inseparable and intimate relationship between humans and nature (see below). This was emphasized for the purpose of political mobilization on behalf of the suffering of the Atrato riverine community as a whole.

When Rio Atrato was given status as a legal subject with corresponding rights that included "rights to protection, conservation, maintenance and restoration" (Judgment T622 2016, 5), the concept of "biocultural rights" was introduced to acknowledge "the deep and intrinsic connection that exists between nature, its resources, and the culture of the ethnic and Indigenous communities that inhabit them, all of which are interdependent with each other and cannot be understood in isolation" (Judgment T622 2016, 3). By recognizing that local communities of different ethnicities have ways of conceiving and relating to the river that are different from how the environment is treated in former legislation, as an object separate from humans and without rights, and by acknowledging the biocultural rights of people and river, the ruling attempts to capture an other-than-modern perspective (or new-modern, given its grounding in liberal law) that is closer to the lived realities of riverine communities. However, although the legal decision does recognize an interdependency between local communities and the environment, it is not very specific in detailing what this entails and for whom. Neither does it cite, describe or in other ways incorporate local concepts, values or laws that reflect the particularities of different riverine communities' relationships to the Atrato.

Along Ganga and Yamuna rivers, the variety of livelihoods and economic activities and spiritual and mundane engagements with the river are again of a different order to those in Aotearoa New Zealand and Colombia. However, in contrast to the other cases where marginal, oppressed or neglected groups are fighting for their own and the river's rights, it is the Hindu majority view that is referred to in the Indian riverine rights ruling. The Uttarakhand High Court declares Ganga and Yamuna Rivers as living entities and legal persons (Mohd. Salim v. State of Uttarakhand Case 2017). The legal document pulls on legal inspiration from different sites, borrowing phrases and elements from the Aotearoa New Zealand case and recognizing the significance of Hindu philosophy. The court decisions state that "Rivers Ganga and Yamuna are central to the existence of half of Indian population and their health and well being" and that Hindus have a "deep spiritual connection with Rivers Ganges & Yamuna" (Mohd. Salim v. State of Uttarakhand Case 2017, art. 17 and art. 11). Other religious and ethnic communities, i.e., Muslims, Christians, Buddhists, Dalits and Indigenous peoples, are not mentioned, and neither is heterogeneity within Hindu beliefs. The court ruling was hurriedly put together and did not include any participatory processes. The ruling has also faced challenges regarding contextual complexities such as size and scale.

Common to all the riverine rights rulings is an attempt to integrate some kind of people-river inseparability not conventionally included in modern state law. Though differently and with varying degrees of success, each ruling attempts to capture the significant relationality that exists between people and river in these different contexts.

Relationality in rights of nature debates

In rights of nature debates, such relationality is often attributed to what is termed "relational ontologies", said to characterize many Indigenous and other locally emplaced peoples' way of thinking about and being in the world (Anderson 2024, Fisher 2022, Muller et al. 2019, Tanasescu 2020). Within relational ontologies, relationships are primary, and entities can only be understood as constituted by relationships to significant others, and not in separation from them (Tanasescu 2020). In contrast, in what is often denominated modern or Western ontologies, entities are primary and are understood to be in dichotomy to other entities. Whereas human-environment relationships within relational ontologies are described as mutually constitutive relationships between a plurality of entities or beings, it is claimed that modern Western ontology is based on binaries such as nature-culture, human-environment, land-water, each of which are understood to be separate phenomena and concepts (Fisher 2022, 18).

Nature's rights cases from India, Aotearoa New Zealand and Colombia (including Australasia and the Americas more broadly) are used as examples of efforts to include relational ontologies within the legal framework and are understood as cases of onto-legal pluralism (see for instance Anderson 2024, Clark et al. 2019, Clark et al. 2024, de Micheaux 2019, Kramm 2020, Macpherson 2023, Tanasescu 2020, 2022). In all three countries, there are examples of rights of nature advocates, community representatives and scholars attempting to describe alternative ontologies of relationality emphasizing the inseparability between people and river. Some have done this in a nuanced and critical way, emphasizing how different ways of conceptualizing rivers and human relationships to them are complexly interwoven in life and law. However, commonly a sharp contrast is made between Indigenous relational ontologies and Western modern ontology, especially within the discourse of rights of nature advocacy. The first of these is commonly romanticized, and blame is placed on the second for creating the problems being addressed by the legal process. Moreover, the term "relational ontologies" is often used to encompass a whole range of vernacular ways of understanding one's relation to nature. Such generalization overshadows the specific ways local and Indigenous people relate to nature and rivers, just as the category Western ontology does not correspond to the lived lives of all conceived of as Western or modern. Here, we will not do an exercise in generalizing and comparing, nor go into the debate of whether the radical difference within the riverine rights cases is best understood in terms of a difference in ontology. Rather, we will investigate specific ethics and practices of relating to rivers in the three cases.

It is evident that a larger turn to *relationality* and *inseparability* and away from *separation* and *dichotomies* is taking place. Posthumanism, feminist materialism, multi-species ethnography and more-than-human approaches argue for acknowledging human embeddedness in the world, the inseparability of human and nature, how the entanglement with others is integral to human existence and how all humans are caught up in intimate relations to other-than-humans. In a growing body of literature on water within social sciences and humanities, concepts such as "waterworlds" (Hastrup 2009), "hydrosocial" (Boelens 2014, Linton and Budds 2014, Krause 2017) and "aguasocial" (Brandshaug 2021) draw attention to the close relationships between humans and water. Here, Indigenous scholarship contributes by showing diverse ways of knowing and caring for water characterized by reciprocity and relationality (O'Donnell et al. 2024, 208). There is an increased interest in conceptualizations of water relationality, in water's own intrinsic relationality and in "thinking relationships through water" (Krause and Strang 2016). These frameworks evade dualistic ways of characterizing human-water relations and turn to specific landscapes and explore what goes on in places. Hence, to avoid an overly dichotomous and generalized understanding, it is helpful to consider the particular ways people relate to their environment in different places. It is important to pick apart what relationality means in each case and reveal both difference and heterogeneity, and what is shared in common.

In the sections that follow, we seek to describe some of the ways people create relationships to their river (and the environment more broadly). This is in line with this book's central endeavour of understanding riverine rights in their context and through encounters with the rivers and their people. Each section is written by different authors with different kinds of relationship to and knowledge of the river in question. Therefore, the content and style vary between them, and each section starts with a statement revealing the positionality of the authors in relation to the river and its people. Although all try to capture something particular about the relationships between Whanganui River, Atrato River, Ganga and Yamuna rivers and their riverine communities, the aspects of people-river relationality emphasized are not the same despite some overlaps. This reflects the diversity of each case and the positionality of the authors. The sections variously rely on photographs, origin stories, ethnographic descriptions, citations from interviews and secondary sources. The aim here is not to make an explicit comparison but rather let the descriptions stand alone and illustrate what people-river relationality or even inseparability means in each context. We hope that the reader will get a sense of the divergences and convergences – of how the cases differ and what could be considered common or shared between them.

Riverine inseparability in place

Whanganui: I am the river and the river is me

By Miriama Cribb

Ko Ruapehu te maunga
Ko Tongariro te mātāpuna
Ko Whanganui te awa
He uri nō Paerangi-i-te-wharetoka
Ko Te Atihaunui a Pāpārangi te iwi
Ko Ngā Paerangi, ko Ngāti Hine o Te Rā ngā hapū
Ko Kaiwhaiki te marae
Ko Rakato te papakainga
Tihei Mauri Ora!
Ruapehu is my mountain
Tongariro is the source of the river
Whanganui is my river
Descended from Paerangi-i-te-wharetoka
Te Atihaunui a Pāpārangi is my tribe
Ngā Paerangi and Ngāti Hine o Te Rā are my sub-tribes
Kaiwhaiki is our place of gathering
Rakato is our ancestral land
Behold, it is life!

I am a born and raised descendant of the Whanganui River, and my connection and positionality begin with a simple retelling of *pepeha*. More than a formulaic tribal expression, *pepeha* is a tool created by our ancestors to frame our existence as Māori, the Indigenous people of the land, through the identification of places and people (Cunningham et al. 2023). This process creates a timeline of who we are, where we come from and where we exist today. More importantly, not only is it important to identify these places and claim them, but to engage with them. So, to say Whanganui is my river also means for my river to know who I am – for my river to see me, hear me and feel me.

From a Māori worldview, the inseparability of people and rivers goes back to the beginning of time, to the origins of place and the origins of people. A Māori worldview acknowledges a natural order to the universe (Davis 2006), where relationships between nature and people manifest through *whakapapa*. *Whakapapa*, often translated as genealogy or descent, is a "system of thought based on an interrelatedness that establishes connections with ancestors and the environment and creates culturally appropriate ways of acting" (Forster 2019, 3).

> *Ko Tahuārangi te waka o Maui-mua*
> *Ko Rangitukutuku te aho*
> *Ko Pikimairawea te matau*
> *Ko Hāhā-te-whenua te ika kei rō wai.*
> Tahuārangi is the waka (canoe) of Maui-mua
> Rangitukutuku is the fishing line
> Pikimairawea is the hook
> Hāhā-te-whenua was the name of the fish while it was still in the water
> (Whanganui Iwi and The Crown 2014)

The origins of the Whanganui River begin when Maui, demigod of Polynesia, hauled up the great fish known as Te-Ika-ā-Maui, the North Island of Aotearoa New Zealand. When the fish broke the surface of the sea, the land and water resounded and Maui's brothers invoked Ranginui, Sky Father, to settle the fish. Ranginui sent Matua te Mana, also known as Mount Ruapehu, the first of the group of mountains to be placed at the centre of the fish. However, Ranginui saw Ruapehu was lonely and placed two teardrops at his feet, one of which became the Whanganui River. As time passed, Ruapehu continued to be lonely and pleaded with Ranginui for company, so he sent companions in the form of four mountains – Tongariro, Taranaki, Ngauruhoe and Pihanga.

The origins of Whanganui *iwi* (tribes of Whanganui) start with the origins of the Whanganui River. There are two primary ancestors of Whanganui *iwi*, Ruatipua and Paerangi-i-te-wharetoka, who were here prior to the arrival of the great migration fleet from Hawaiki, the ancient homeland of the Māori. Both Ruatipua and Paerangi named places and landmarks in the Whanganui region, and their *mana* (prestige, authority) is etched in our collective tribal memory, which continue to this day. I share these narratives, sourced from the Whanganui River Deed of Settlement (Whanganui iwi and The Crown 2014), which reflect the commitment of the Crown to address grievances of Whanganui *iwi* in relation to the Whanganui River. I make these observations knowing there are variances within our stories and wish to acknowledge all tribal knowledges and histories.

> And so we go back to the river, and the river is the beginning, the beginning for our people from the mountain to the sea. It ties us together like the umbilical cord of the unborn child. Without that, it dies. Without that strand of life, it has no meaning. The river is ultimately our *mana*. Our *tapu* (sacredness), our *ihi* (force), our *wehi* (awe), all these things make up what the river means to us. It is our life cord, not just because it's water – but because it's sacred water to us.
>
> (Quote from tribal elder the late Matiu Mareikura, Waitangi Tribunal 1999)

For Whanganui *iwi*, the interrelated ancestral ties bind the people and the river in what tribal leader Te Tiwha Puketapu describes as *Rarangi Matua*, or the chronological ancestral sequence connecting the celestial and temporal realms (Waitangi Tribunal 1999). The success of Te Awa Tupua Act can be attributed to the ontology of the river as a spiritual ancestor of Whanganui *iwi* (Kramm 2020). As such, a relationship expressed through *whakapapa* considers the holistic totality of the relationship between humans and nature and the inextricable links between Whanganui River and its people. These links are pertinent to understanding Te Awa Tupua, which is best summarized in *Tupua te Kawa*, the natural law system for the Whanganui River. Te Awa Tupua Act not only acknowledges the nature of the river in a Western construct but reasserts and restores how we have always understood our river, as expressed in *Tupua te Kawa* (Tinirau et al. 2020).

Tupua te Kawa explains four universal truths or principles that convey our relationship as Indigenous people with the Whanganui River (Te Pou Tupua 2023). These laws are:

- *Ko te awa te mātāpuna o te ora* – The river is the source of spiritual and physical sustenance.
- *E rere kau mai te awa nui, mai te Kahui Māunga ki Tangaroa* – The great river flows from the mountains to the sea.
- *Ko au te awa, ko te awa ko au* – I am the river and the river is me.
- *Ngā manga iti, ngā manga nui e honohono kau ana, ka tupu hei awa, hei awa tupua* – The small and large streams that flow into one another form one river.

The third law specifically speaks to the human-river relationship between Whanganui descendants and the Whanganui River, which upholds our inalienable connection and responsibility to our river, reflecting the relational ontology of *whakapapa*. In its simplest terms, I am the river and the river is me is stating that we, its descendants, and the river are one and the same (Allen and Tinirau 2021). The beauty of applying *kawa*, defined by Tinirau et al. (2020) as the values derived from the celestial realm that have permanence, is that they are timeless, and their application and expression will be diverse and defined by each generation of Whanganui descendants. We are still guided by traditional teachings, handed down as tribal epistemologies embedded in *whakapapa, pepeha, whakataukī* (proverbs) and *tikanga* (practices, customs) and maintained and actioned through tribal *wānanga* (forums), *kaupapa* (philosophy), strategies, plans and programmes (Cribb et al. 2022).

Today, for me and many families who have a long inhabitation of our river, our connection and engagement with our river is manifested in many ways. We transmit and uphold knowledge about our river through songs and stories, embedded in education curricular and tribal forums. We maintain the river's identity by ensuring our language, including our tribal dialect, is

Figure 5.2. A canoe on the river. Every year Whanganui descendants arrange a journey down the river from the mountain to the sea, *te Tira Hoe Waka*. Photo by Ellen Hofsvang.

upheld and thriving. We honour and advocate for our river by fighting for its effective management through political involvement and representation in council offices, board rooms and various governing bodies. We exercise guardianship and reciprocity by taking care of our river with river clean ups or give thanks by reciting incantations at the river before and after significant events. We replicate the importance of the river's health and wellbeing through recreation, water sports and by holding events on the river. We protect its spiritual significance through customary practices such as *rāhui*, which restrict and regulate human activity.

Rāhui is a temporary ritual prohibition that restricts access to and bans the harvesting of resources in a particular area. A *rāhui* is put in place as a conservation measure or as a means of social and political control of an area for a variety of reasons, including pollution by *tapu*. The Whanganui River has experienced deaths in and on the river, and placing a *rāhui* is common practice when someone goes missing, drowns or dies in or around the river. Death pollutes water and the things living within it, and a *rāhui* is a practice that separates the living from things *tapu* until such time the river is rebalanced and replenished. After an agreed lapse of time, the *rāhui* is lifted. A *tohu* or sign symbolizing the placement and lifting of the *rāhui* can be the placing and natural removal of a wooden stick or post in the ground, or the absence and presence of natural features such as fish, river species, nearby plants and trees, or birds. A *rāhui* can be initiated by an elder, someone of rank or a tribal expert and is placed and lifted with appropriate rituals and incantations.

To have a relationship with our river means to know and feel the river in a relational way, a view where all things are connected, not existing independently or in isolation. During our project, we were fortunate to visit Kaiwhaiki Marae, one of many villages along the Whanganui River, and speak to a group of local women. They provided lots of examples of how the river was important in their everyday lives, including that we, the descendants, are the voices for the river. The river is our friend and our foe; it has a life force of its own, a spirit. It is our teacher; we learn how to swim, and we know and learn its movements with the tides, floods and the moon. The river has its own eco-system – its tributaries and the swamps, creeks and plant life. When the river is well, the people are well, and when the river is sick, the people are sick. The women spoke of the harmful and damaging relationships, of how dirty the river is, the damming of the river, the decimation of fish supplies, and waste dumped into the river from industrial sites.

While the passing of Te Awa Tupua Act restores and reasserts how we have always understood and had a relationship with our river, it is not without its challenges, especially in terms of conceptualizing the relationship through the legal device of "personhood". As shared by some of the women we spoke to at Kaiwhaiki:

> We are also interested in the term 'personhood' as referred to the river. It is a 'foreign' '*pakeha*' [Māori term for a New Zealander, usually of European descent] concept. You can't eat out of the river and then call it a 'person-hood'.

In response to the above quote, another woman commented that of course the river is not a person nor is personhood a term we embrace, but it is a mechanism put in place so that others will understand that you must treat the river with the respect and agency that an ordinary person has. So, whilst the river is an ancestor to us, this is not necessarily a human ancestor but a greater natural being with its own life-force, energy and spirit whom we must care for and protect. This protection, now to be upheld in Western law, is practised and manifested in how we see, hear, feel, think and engage with our river as exemplified throughout this chapter. More importantly, it is the embodiment of "I am the river, and the river is me". For we are reminded of this responsibility through another tribal proverb: *Kauaka e kōrero mō te awa, engari kōrero ki te awa*! "Don't merely talk about the river, rather speak to the river!" (Wilson 2010).

Dodromá: Our majestic Atrato River

By Jenny Moreno Socha, Daniel Alfonso León and Arlinton Cuesta Mosquera

We would like to recognize the diversity of experiences within the communities that inhabit the Atrato. From river guardians to academics and officials, each voice brings a unique perspective on the relationship with the river. It is important that the voices of those who have been affected by mining exploitation are heard, in order to build a more complete understanding of the situation.

The following writing is based on the compilation of stories from interviews conducted in the field within the framework of the Riverine Rights project, by researchers from the Colombia team. One of us grew up and lived along the Atrato River, while two of us are PhD students from other parts of Colombia who have come to know the river and its people through fieldwork in the area. The voices that are captured below are from Indigenous and Afro-descendant communities, who have different roles: Guardians of the Atrato River, Indigenous authorities and academics or government officials, who behind their professional investiture are also people who were born and raised on the banks of the Atrato River. Therefore, despite the fact that they represent certain institutions, when they refer to the relationship between communities and the river, they express their personal experience. In addition to this, voices are also captured from national officials and members of academia or NGOs, who, following the ruling T-622 of 2016, began to develop relations with the Atrato River. These voices are also important to recount because they are a sign of the commitment and importance that the ruling has had in the different instances and, above all, the connection that the river generates with these people, who began to mix with the daily dynamics of the communities and the river.

The majestic Atrato River meanders through one of the most biodiverse regions on the planet. It is a giant that sustains the lives of Black Indigenous and mestizo populations along its course, until it flows into the Gulf of Urabá on the Caribbean Sea. Its waters embrace the landscape, transforming into other rivers that beat to the rhythm of chirimías, bullerengues, bundes, pasillos, bambazú and danzón – rhythms of Afro-Colombian music.

The Atrato River is not just a stream of water; it is a vital link, a constant companion, a teacher and a refuge for the communities that live on its banks. Here there is no division between the river and those who live along it; life itself flows with its waters, carrying with it stories, knowledge and hopes.

"The river is our blood, it is the path that connects us to the world and to ourselves", says a woman from the community. This expression encapsulates the inseparability of the Atrato and its people, who find in its waters not only sustenance but also identity and a sense of belonging.

The river speaks in whispers and roars, sometimes calm, sometimes furious. Its cycles mark the rhythm of everyday life: "When the river rises, so do our hopes, but so do our worries. When it falls, it seems to take our anxieties with it, leaving behind a temporary relief", shares a fisherman as he mends his net.

In the community's songs and poems, the river appears as an omnipresent, almost human character, described as a mother, a home for humans and fish alike, and a place of memories from the past and dreams for the future.

Everyday practices also reveal this symbolic relationship. In the early morning, women walk to the banks to wash clothes while singing verses that tell ancient stories. Children play barefoot on the banks, learning from their elders the secrets of water: how to read its currents, how to respect its whims.

"The river teaches us patience and humility", says one community leader. "When it gives, we are grateful; when it takes, we also learn". This balance of gratitude and resignation underlines a philosophy of life rooted in reciprocity. The river provides fish, water for drinking and irrigating crops, but it can also reclaim land with its floods. Communities, in response, adapt their customs, building houses on stilts or temporarily migrating when the river demands it.

The Atrato is also a memory. An old man tells how its waters carried not only goods but also news and dreams. "The river was always the messenger. In its canoes went the joys and sorrows of all of us". Even today, stories from the past flow with it, reminding new generations that the river is not only a landscape but living history.

In spiritual ceremonies, the Atrato plays a central role. "It is a sacred being, a spirit that listens to us", says a local shaman. In rituals of purification and gratitude the river is depicted as a living entity with its own soul, deserving of respect and care.

This inseparable relationship is also expressed in the values that the river inspires: solidarity, resilience and harmony with nature. "When the river suffers, we all suffer. That's why we fight to protect it", says a woman as she watches a sunset that turns the waters golden. These words echo a deep commitment, a collective responsibility that transcends generations

The Atrato and its people are not understood as separate entities. They are a unique fabric, a single story told in water and land, in hearts and currents. As a local proverb says: "Whoever listens to the river, listens to the voice of his own soul".

Figure 5.3. Fishing on the Atrato River. Photo by Daniel Alfonso León.

In this way, the Atrato River is loaded with history and collective memory, reflecting a part of the history of Colombia. In the words of Manuel, a local man and researcher in environmental toxicology from Chocó:

> If we look at the history of Colombia and Chocó, it says that people have always been created and created around rivers, right? Rivers are always sources of life and many other things, and these populations are always close by, but that is a mechanism of subsistence because the river provides food, transportation, and many other things that we need to live.

It can be said that the history of the Atrato River begins with the ancestral relationship it has with Indigenous communities. For the Embera community of Chocó, the river is the heart of their culture. In their worldview, there is a law of origin where, in the words of the Indigenous representative Plácido:

> All nature has its own right, the river has its own right, the territory has its own right, the forests that surround the territory [...] and that right many times, well, if that law of origin is not known from that worldview, then the rights of the rivers, of the territories, of nature are violated and it is there where there is a difference quite far from the Indigenous peoples compared to the non-Indigenous society when they ignore, violate and mistreat.

For them, the Atrato River is historically known as Dodromá. Germán, an Indigenous leader, explains:

> 'Do' means river and 'droma' means big, so: 'big river', and that is why we, the old people, always say, I am going to 'Dodroma', I am going to the big river, because the Atrato is a place of fishing, a place of rest, a place of exchange, a place to travel. That is how we have always considered it.

But the river is also nourished and grows through its relationship with Afro-descendant communities. For these communities, the Atrato River is presented as a continuous and sustained story over time. This river contributes to explaining events, stories and a chain of events that give meaning to their experiences. In these senses, the river is dynamic and biodiverse. Giovany, who knows the river as a researcher but also through growing up along it, says:

> I am from Atrato and our feelings about the river are profound. We believe that the Atrato is the spirit of the people of the territory and that when people do not have spirit, they are dead. So, for us, if the Atrato disappears, so does the magic of the Atrato, the dynamics of the Atrato, that connection between the people, the river and the biodiversity of the river. It is not only the mirror of water, the Atrato River is the one that carries the rise of the Bocachico, it is the one that allows and was the first navigation route of the territory. We go out through the Atrato both to the Pacific and to the Caribbean, so it is a matter of the spirit of the condition of the connection of the people with nature and with the territory, that is the Atrato.

Harry, another local who works as a professional on issues of sustainable development in the area, describes what he calls "a river culture" that is shared between Afro-Colombians and Indigenous people living along Atrato:

> We have always been linked to the Atrato River. In fact, we have a custom, there is a river culture; for blacks, for Indigenous people, there are different ways of locating themselves, but we have a river culture. The Indigenous people are always in the upper parts of the rivers, but we blacks are in an area where we can communicate, that is, the river is not necessarily useful for communicating, for moving, for bringing products, and if you look at any large black community, it is on the banks of a river.

However, this majestic river and the communities that live and inhabit it are not alien to the complex reality that plagues Colombian territory: the impact on rivers caused by the extraction of minerals and wood and the impacts of the armed conflict. One of the river guardians, Alexander, describes this in poetic terms:

> I remember that the Atrato was a river of mystery
> and today the treatment it receives is that of a pure cemetery.
> It is not only the mining that has contaminated the river,
> it is also human bodies that have been thrown into that river
> things that men do, which generate consequences
> humanity ends with the damned violence.

Contemplating such feelings leads us to think about the complex relationships woven around the river, as well as the vulnerability of populations to the effects of the environmental impacts of mining and deforestation. Liliana, part of the expert panel created after the T-622 ruling, looks at the river with despair because it is dying, as people say. Manuel again, expresses:

> When I look at the market and see all those little fruits and colours, it means life to me, that's what it means to me, because everything the river generates is very important for the communities: life, food, transportation, economy and production. It's impressive, that's why it hurts me so much to see it being destroyed. For us, it's very painful to see how these habitats that were once spectacular and still are being destroyed.

Despite the river facing a dynamic of change forced by the logic of logging, mining and other conflicts that tend to exacerbate local needs and problems, Indigenous and Afro-descendant communities continue to resist and try to preserve the river in the face of great challenges. Manuel further reflects on this:

> I think that this is because these people preserve their ancestry, right? They preserve all those characteristics that they have had from generation to generation: their rituals, their food, their religion, all their ancestry, all those customs that they have carried out and that they have carried out throughout their lives and their generations can be preserved, because many of these activities are lost when mining activities begin to take place and many people from other places begin to arrive. That means that, to a certain extent, this ancestral knowledge is being lost due to more contemporary or more urban activities, so to speak, so, we must also protect that, we must try to preserve these ancestral customs and continue to carry them out.

But they are not the only actors interested in the preservation and recovery of the river. Ruling T-622 of 2016, which declared the Atrato River a subject of rights, allowed the problems of the Atrato to become visible – it brought new eyes to the river but above all brought about new feelings and relationships between these institutional, academic and NGO actors with the Atrato River, and through their role they try to protect the river:

I am grateful for the river and its people because it has strengthened me in the role I have. Going to the territory, not only at the desk but in the territory, and learning with them, having a drink, [...] I have had a very nice experience with the people above all, and of course, with the contribution that I can make from my role to improve from an environmental point of view, so that one can have better tools and a stronger voice.

(Fabian, government river guardian)

For me, rivers are my veins [...] so I work to bring them back to life, that is my philosophy, my essence, my reason for being in life; I have understood that my public service, my personal and professional capacity is to do something for those who need it, and for me everything is the sum of a river, whatever it may be, with its tributaries and its basin, its territory, the soil, the flora, the fauna, all the forms of life that sustain that basin and its people, that is the sum. Yes?

(Martha, government official)

By dedicating time and effort and also getting to know the guardians and the people there, you certainly start to build an emotional connection with the river itself. And above all with what the river represented before; I have heard many people talk about the river before, about how everything was open-door. So, you could simply go to your neighbour's house and go in, you could sleep peacefully outside without any problem, there was plenty of food, there was a party, people could walk along their river whenever they wanted, and there was this feeling of freedom that you notice in the way they speak and a kind of nostalgia for what has been lost because of all these things that have been associated with it in the last 30 years.

And this work is a bit about that, like that image that people have when they remember the river as it was in the 80s, early 90s, and for me it also connects emotionally because I am from Villavicencio; I am from Meta and I also have a special connection with a river called the Yucao River [...] and I go to the Atrato and I see the Yucao too. Fortunately, the Yucao has not had these problems that the Atrato has, because there are territorial and historical differences, but I always think of the Yucao when I see the Atrato and like the smell of the rivers, that smell that they have, right? So, what does it mean to me? Doing everything possible with the people there to return to that old Atrato or to bring the Atrato of today as close as possible to that old Atrato.

(Rodrigo, socio-legal researcher)

Divine rivers: Ganga and Yamuna

On a shelf in his living room stands a small glass bottle of water. The water is Ganga water and comes from the great river that runs through large parts of India. Just as the Ganga River is a holy divinity for Hindus across India,

> Ganga water is considered sacred. Bibhu grew up and lives far away from Ganga, yet, as an Indian Hindu, the river is spiritually important to him. The significance of the river and its waters is manifested in this little bottle of water that puts him in a relationship to the river even though it flows far from where he lives. Bibhu tells us that Ganga's sacred water is used in Hindu purification rituals across all of India.

The description above is a rendering of something our Indian colleague Bibhu Prasad Nayak told us in a project workshop. Unfortunately, Bibhu and the other researchers on the project's India team could not, for different reasons, co-write this chapter, but they have contributed by reading through and commenting on chapter drafts. Therefore, this chapter section on India is written based on written secondary sources, rather than ethnographic fieldwork with riverine communities or first-hand experiences from Ganga and Yamuna rivers, as is the case for Whanganui River and Atrato River in the previous sections.

In common with Bibhu, Hindus across all of India have a religious relationship to Ganga. Additionally, those who live along Ganga and Yamuna – both Hindus and non-Hindus – engage with and relate to the rivers in myriad ways. This section focuses on people-river relationships in India and the significance of these rivers.

Ganga River and Yamuna River are Hindu religious entities. They are both considered holy rivers that "form the core spiritual sources and religious identity of Hindus" (Ranjan 2024, 6). Ganga is personified as "Ma Ganga" or "Ganga Mata" (Mother Ganga) and "Ganga Devi" (Goddess Ganga) and is the origin of all of India's sacred waters (Eck 2012, chap. 4). She is also referred to as "the embodied goddess of the gods", said to derive from heaven, from Vishnu's foot, and was carried in Brahma's water pot, and then she is said to have flown from Shiva's head and down to earth (Eck 2012, chap. 4). Hindus bathe in Ganga, make offerings to ancestors and gods with its water, shed the ashes of dead relatives, flowers and earthen lamps into the river and bring bottles of Ganga water home. "The Ganga is holy all along its course, from its source to the sea" (Eck 2012, chap. 4). Likewise, its water is a sacred and purifying liquid that is used not only along the banks of the river and in temples but also in homes, as the introductory vignette illustrates.

Like Ganga, Yamuna River is considered to be of celestial origin, the daughter of the sun, who fell on mountain Kalinda and came down to earth. Kalinda's glacier is today seen as Yamuna's origin source. From there its water gathers in lake Saptarishi Kund and flows to Yamunotri, where it blends with hot spring water before it flows further through Uttarakhand, Delhi and onwards, ultimately merging with Ganga River. There are several sacred spots along the way where the goddess "Ma Yamuna" (Mother Yamuna) is revered by pilgrims, through images, bathing and songs. Like with Ganga, her waters are purifying but also "the waters of love" (Eck 2012, chap. 4).

The significance of these divine rivers for Hindus is not only restricted to religious myths. "Throughout history, rivers have been revered as a means for individuals to connect with the divine and achieve a sense of unity" (Ranjan 2024, 6). Moreover, Hindus today engage in significant relationships with these rivers. de Micheaux (2019) gives several examples from written sources and from her own interviews regarding Hindu perceptions of "an internal and essential link between people and the river" (p. 197). Furthermore, Ganga, especially, is closely tied to people's identity (de Micheaux 2019, Ranjan 2024).

Alley describes how residents of Varanasi, by the riverbanks of Ganga, see the river today; "they view Ganga as a Mother [...] whose flow provides for everyone. Devotees feel related to her as a child to a Mother, and they seek protection from her. Devotion and worship rituals are critical to ensuring that protection" (2019, 9). Moreover, pilgrimage and other religious practices are quite prevalent in India. Hence, Ganga's religious significance, as well as Yamuna's, is not limited to belief, imagination or conceptualization of the river as divine but is also reflected in bodily and emotional interactions through pilgrimage, songs or "everyday interactions [that] are not restricted to riverbank residents but apply to any Hindu devotees who perform prayers, songs, or rituals in relation to the River Ganges" (de Micheaux 2019, 201). Religious river practices are seen as ways of communication between human individuals and a divinity (Alley 2000).

Another side to the religious perception of Ganga River is its role in politization processes (Alley 2002, Mawdsley 2005, de Micheaux 2019). The sacred Ganga River has been politicized for different purposes and with different degrees of success; for instance, to strengthen Hindu nationalism and in anti-dam struggles (de Micheaux 2019, 198).

Looking into the political ontological dimensions of human-river relations in anti-dam movements, de Micheaux renders how the religious Ganga River is pictured in anti-dam religious ontology:

> In what may be called the 'Hindu religious perspective', the river is both a material river and a Goddess. The anti-dam 'religious perspective' relies on an ontology which does not consider the river as an object. Furthermore, the perspective does not neatly separate humans from the river. As stated by an anti-dam activist, the relation to the river involves questions about 'who we are', or as stated by Agrawal, 'our very identity' [...]. With its purifying power, the river transforms humans.
>
> (2019, 200)

Others, on the other hand, see the river as a material entity, a resource that is external to humans and can therefore be exploited. Although this view is primary held by those advocating for hydropower, more scientifically oriented parts of the anti-dam movement perceive the river as a separate, material entity and distance themselves from the religious conceptualization of the river. Others again combine a religious and a scientific-ecological valuation of

Figure 5.4. Hindu pilgrims immersing themselves in the sacred water of the Ganga. Photo by Axel Borchgrevink.

the river in their environmental activism (de Micheaux 2019, 200). In the conclusion, de Micheaux (2019) points to the Uttarakhand court decision of giving rights to Ganga and Yamuna rivers as an example of implementing an alternative river ontology in the legal framework.

In another vein, Ranjan discusses religious environmentalism in relation to the river activism of Saddhus (religious ascetics) and riverine rights in India. The Saddhus he interviewed call the river Ganga Ma out of respect and view her as an honourable sacred deity, religiously put above humans. When supporting the act of giving Ganga rights as an important means in the fight against environmental issues related to the river, however, "[t]hey do not believe assigning a sacred status serves any purpose in the debate on the river's rights. Quite surprisingly, they instead argued that river must be treated [equally] as humans" (2024, 15). Hence, they argued against a pure transmission or translation of Hindu values of the river into political-juridical interventions and instead thought the river should have the same status as humans in law, a view that seems to accord with how they see the river themselves. Likewise, those Alley (2019, 9) interviewed also expressed that the concept of a legal person or just person did not resonate with how they see Ganga as a female God.

The decision of Uttarakhand High Court to accord legal personhood to the Ganga and Yamuna was created in response to two public interest litigations regarding encroachments and high levels of pollution in the Ganga. These rivers are contaminated entities. Having hundreds of millions of people living on their banks, in addition to pilgrim tourism, damming for large-scale hydropower-project, mining of riverbed material and extensive infrastructure projects on or by the river, issues of waste and pollution increase, with the general negative environmental impact (Eck 2012, Ranjan 2024) coinciding with more floods, landslides and other climate change disasters (Upadhyay and Nayak 2024, 15, Borchgrevink and Upadhyay this volume).

In her book titled *On the Banks of the Ganga: When Wastewater Meets a Sacred River*, Alley (2002) addresses the apparent conflicting relationship between Ganga as contaminated and Ganga as sacred, emphasizing the paradoxes and complexities concerning Ganga. Hence, as Eck points out, "(t)he health and vibrancy of the Ganga and India's other rivers is both an environmental and a theological issue" (2012, chap. 4).

Additionally, Ganga and other rivers of India are troubled and politicized in the sense that they are imbued with caste structures, as is the environment more generally in an Indian context (Prasad 2022, Sharma 2022), and land in particular (Nielsen et al. 2020). This has implications for Dalits – formerly known as untouchables – and the way they relate differently to these rivers. Several authors writing from an Indian context have pointed to Dalits and their environmental struggles, the relationship between caste and socio-environmental justice, how environmental problems are differently distributed and how marginalized groups such as Dalits, Muslims and Indigenous people embody environmental problems.

Dalits experience many forms of oppression and discrimination, among them limited access to land and water (Prasad 2022, Sharma 2019, 2022). The environmental experiences and everyday practices of Dalits are shaped by domination, exclusion, and restrictions; "structured by an archaeology of untouchability in body, contact, touch, smell, feel, belonging, work, and sociability" (Sharma 2022, 88). As with other commons, water bodies are often sites of "social conflicts and political protests, domination and resistance, construction and destruction, exclusion and violence" (Sharma 2022, 87). Moreover, water is tied to specific ideas of purity and impurity, where a Dalit's contact with water is seen as contaminating (Sharma 2022, 84).

In Dalit narratives, their relationships to Ganga and Yamuna, and rivers more generally, are described in rich ways. In Dalit literary text their relationship to nature is described in varied and detailed ways that goes beyond mere spiritual relationships and thus differs from mainstream environmental literature often inspired by Hindu religious imagination of nature (Sharma 2019). In these text, Ganga and Yamuna appear frequently. In a novel by Dalit writer Adwaita Mallabaran, the river Titash is described in detail – its fish, fishing and life along and in the rivers – highlighting subsistence work, cultural activities and painful human experiences. Sharma cites the following story:

> Titash is the name of a river. Its banks brim with water, its surface is alive with ripples, its heart exuberant. It flows in the rhythm of a dream … But all the rivers are not similar in appearance and essence. How they behave with people differs and how people behave with them differs too … The river there is a frenzied sculptor at work, destroying and creating restlessly in crazed joy, riding the high-flying swing of fearsome energy … Titash is just an ordinary river. No one will find its name in the history books, in any of the chronicles of national upheavals … But its banks are imprinted with stories of a mother's affection, a brother's love, the caring of a wife, a sister and a daughter. That history perhaps some know, perhaps some don't. Still, that history is true and real.
>
> (Mallabarman 1993 [1956], 10, 26, in Sharma 2019, 1018)

> The relationship between people, river, and seasons permeates events and images, metaphors and musings, songs and dialogues … Daily life and festivals, events and reveries reflect the keenly perceived changes in nature. The river both evokes and affects lives and thoughts.
>
> (Mallabarman 1993 [1956], 4, in Sharma 2019, 1019)

The text clearly expresses a close relationship between people and river. This is apparent in many varied ways in Dalit environmental literature. Sharma explains that

> [a] river is not just a sacred, pure, religious and spiritual body; it is not only a place for devotees and their prayers. The river also reminds Dalits of their pain, body blows, the struggles of boatmen, waves causing havoc to boats, and struggles for survival.
>
> (2019, 1023)

He also highlights "how Dalits longed to change their landscape in which they had to live in poverty and drudgery into a liveable place" (2019, 1024), exemplified by a passage in Siddalingaiah's autobiography:

> A big drain lay in front of our house. It was the holy Ganga to us. It would fill up and flow during the rainy season. Watching the flowing drain was great fun. A variety of objects came floating by. Coloured paper, clothes, old cradles, baskets, palm fronds, fish and snakes float by. Once a corpse came floating … Even when it stank, our affection for it did not wane. […]
>
> People from the factory on the other side dumped mounds of black earth into it. Small pieces of bronze and copper could be found in these mounds. We would descend into the drain and gather them by digging up the mounds … For us the drain had become an inexhaustible trove that could earn us candy … We hoped the drain could throw up even more

> precious things. So we would descend into it and walk for a long time. As it was a sewage line, it would be like a tunnel journey.
>
> (Siddalingaiah 2003, 29, in Sharma 2019, 1024)

As opposed to what has been highlighted in the beginning of this section about mainstream Hindu imaginaries of Ganga, in Dalit literary work, the environment is not depicted as consisting of entities superior to them but rather a place of "wonder and fear" (Sharma 2019, 1025). Moreover, the rivers Ganga and Yamuna are not sacred for all who related to them.

Concluding remarks

The relationship that people have with rivers varies within and between cases. In all of the cases considered above, it can, however, be commonly described as one of inseparability, although this does not entail the same for all. There are examples of this inseparable relationship being productive and positive, described in terms of reciprocity, care, love and strength, and being of a more harmful character, related to the powers of the river, struggles for survival, destructive activities and the negative ecological state of the rivers that affect those who are close to it. The rivers are described as entities with knowledge, stories and memory, someone to speak with, to listen to and to protect and be responsible towards. Moreover, relationships of hope, wonder and solitude but also danger and fear are revealed in the descriptions above.

The emphasis on relationality is also key in all three cases, as is common for people with strong connections to places. In each of these three cases human relationships to the river in question are markedly different from the hegemonic understandings of human-environment relations characteristic of modernity, the state and formal legal frameworks. This difference can be framed as one between separation and inseparability. Therefore, there is a shared novelty in the rulings and laws on river personhood in that they try to transcend what has dominated legal thought in (post)colonial settings. The legal decisions attempt to include understandings of people-river inseparability, despite challenges of translation and incompatibility. This is successful to very different degrees in the three cases, covering a spectrum from partial mention to expressed decolonial effort.

While personhood status has extensively been emphasized as the novelty of these legal innovations, especially in media coverage and rights of nature advocacy, the idea of river personhood only resonates with local ways of conceiving of and relating to these rivers to limited degrees.[1] Contrary to the common perception that the personhood concept reflects local conceptualizations of the rivers, in none of the cases is the concept a reflection of what the river really is to local riverine communities. As conveyed in the descriptions above, the rivers are variously described as a living being or entity, ancestor, spirit, divinity, friend or teacher – terms that can be attributed to or confused with that of a person but that still do not mean the same.

However, the cases demonstrate that there are other ways – through different kinds of concepts and principles – in which the riverine communities' relationships to the rivers are described in the legal documents.

Whanganui *iwi* notions are substantially incorporated in the Te Awa Tupua Act, acknowledging and describing the particularities of their special relationship to the Whanganui River and to their local environment more broadly. Here, its tributaries, mountains, sea, land, ancestors and kin are included. The thoroughness in encompassing Whanganui *iwi* values is part of decolonial efforts after centuries of injustices and oppression that have included, in particular, the dispossession of their land and waters. It is a result of a long settlement process and of Māori participation in drafting the Act.

In the Atrato case, the concept of biocultural rights is coined to express the interconnection of humans with the environment and to reflect how local communities relate to Atrato. Moreover, the significance of Atrato River for different local groups is thoroughly acknowledged in the ruling T-622, albeit not very clearly detailed or differentiated.

In the legal documents of Uttarakhand High Court, the spiritual significance of Ganga and Yamuna is referred to, and the concept legal person is sidelined with living entity. A common critique from Hindus has been that divinities like Ganga or Yamuna cannot be reduced to the legal concept of personhood. And, to reflect in the legal documents what the Ganga and Yamuna are to Hindus is not necessarily conceived of as the best way to secure Ganga's legal protection and health. Moreover, the emphasis on Hindu spiritual connections with the river and a larger tendency of religious environmentalism in India disregard other kinds of relationships to the river, in particular marginal groups' embodiment of environmental issues. Critics suggest that riverine rights in India are a form of "Hindutva lite" (Alley 2019), i.e., a legal expression that strengthens a dominant Hindu nationalist ideology (Borchgrevink this volume).

Hence, while the inclusion of locally specific people-river relationships in legal texts in Colombia and Aotearoa New Zealand are locally perceived as having positive potential for the wellbeing of people and rivers (although not without its challenges), the India example is not quite so clear. In all the cases, there are evidently also potential pitfalls and challenges regarding essentialization, exploitation and reproduction of asymmetrical power relations (see also Brandshaug and McNeish this volume). Apart from the question of what is included in terms of a people-river relationship, the how and who of this also becomes important.

In the case of the Whanganui River, the significance of the Māori involvement through the long process of a negotiated settlement (Cribb et al. 2024, Cribb this volume, Borchgrevink this volume) cannot be overstated as vital for the apparent success of the Te Awa Tupua Act. Moreover, a variety of actors with interests in the river, also including non-Māori, have been involved alongside Māori. The process has been an exercise in weaving together ancestral Māori and modernist conceptions of relating to rivers and doing law, both of which many Māori are experienced in deploying in their everyday life (Salmond 2014). When it comes to this interweaving and the reflection of Whanganui *iwi* concepts of people-river inseparability in the Act more specifically, participation stands out again as important for the degree of success in reflecting and safeguarding locally specific relationships.

The Atrato ruling resulted from the initiative of local communities to address the health risks posed by the contaminated river, but the actual representation of local communities in the drafting of the ruling was not significant (Vallejo Piedrahíta, McNeish, Moreno Socha, Cuesta Mosquera and Alfonso León this volume). Even more so in India, where the ruling was very much a top-down decision made by the court in a hasty manner without any participation in the process by river communities (Borchgrevink and Upadhyay this volume). Without any participatory processes, the particularities of and diversity in people's ways of relating to their rivers could only be included to a very limited degree.

In this chapter, we have sought to describe the particularities of people-river inseparability along Whanganui River, Atrato River, Yamuna and Ganga rivers, while recognizing that our descriptions are only partial depictions of the heterogeneity and complexity in how people relate to these rivers. The complexities and nuances of such emplaced relationships testify to the difficult task of safeguarding them through legal decisions. Moreover, our case studies suggest that local representation and extensive participation is an essential requirement for any process aiming to give substantive attention to and understanding of local particularities. This is especially the case when the intention is to do law differently and ensure legal pluralism that better safeguards both rivers and their riverine communities together.

Note

1 For similar discussions on the personhood concept, see Alley (2019), Clark et al. (2019), Kramm (2020), Rajan (2024), Strang (2020) and Tanasescu (2022).

References

Alley, Kelly D. 2000. "Separate domains: Hinduism, politics and environmental pollution." In *Hinduism and Ecology*, edited by Christopher Key Chapple and Mary Evelyn Tucker. Harvard University Press.

Alley, Kelly D. 2002. *On the Banks of the Gaṅgā: When wastewater meets a sacred river*. University of Michigan Press. doi:10.3998/mpub.12072.

Alley, Kelly D. 2019. "River goddesses, personhood and rights of nature: Implications for spiritual ecology." *Religions* 10 (502): 1–17. doi:10.3390/rel10090502.

Allen, Ana, and Rāwiri Tinirau. 2021. *Te Oranga O Te Awa Tupua: Supplementary Report - aspirations and priorities of Te Awa Tupua Uri*. Te Atawhai o Te Ao.

Anderson, Mark. 2024. *The rights of nature and the testimony of things*. Vanderbilt University Press.

Boelens, Rutgerd. 2014. "Cultural politics and the hydrosocial cycle: Water, power and identity in the Andean Highlands." *Geoforum* 57: 234–247. doi:10.1016/j.geoforum.2013.02.008.

Brandshaug, Malene K. 2021. "Water, life, and loss: Aguasociality and environmental change in the Peruvian Andes." *Kritisk etnografi – Swedish Journal of Anthropology* 4 (2). http://urn.kb.se/resolve?urn=urn:nbn:se:uu:diva-463873.

Clark, Cristy, Nia Emmanouil, John Page and Alessandro Pelizzon. 2019. "Can you hear the rivers sing? Legal personhood, ontology, and the nitty-gritty of governance." *Ecology Law Quarterly* 45 (4): 787–844.

Clark, Cristy, Karen Fisher and Elisabeth Macpherson. 2024. "Indigenous rights and ontological plurality in the institutional arrangements for the Waikato and Waipā Rivers in Aotearoa." *The International Journal of Human Rights*: 1–28. doi:10.1080/13642987.2024.2361650.

Cribb, Miriama, Elizabeth Macpherson and Axel Borchgrevink. 2024. "Beyond legal personhood for the Whanganui River: Collaboration and pluralism in implementing the Te Awa Tupua Act." *The International Journal of Human Rights*: 1–24. doi:10.1080/13642987.2024.2314532.

Cribb, Miriama, Jason P. Mika and Sarah Leberman. 2022. "Te Pā Auroa Nā Te Awa Tupua: The new (but old) consciousness needed to implement Indigenous frameworks in non-Indigenous organisations." *AlterNative: An International Journal of Indigenous Peoples* 18 (4): 566–575. doi:10.1177/11771801221123335.

Cunningham, Chelsea, Anne-Marie Jackson and Hauiti Hakopa. 2023. "Walking our pepeha." *MAI Journal* 12 (1): 15–24. doi:10.20507/MAIJournal.2023.12.1.2.

Davis, Sharlene, ed. 2006. *Ka Whakaoho Ake I Nga Tangata Nunui: Māori economic development – Awakening the sleeping giant*. Reed Publishing.

de Micheaux, Flore L. 2019. "Politicizations of a sacred river: The story of Gaumukh-Uttarkashi eco-sensitive zone, Uttarakhand, India." *Journal for the Study of Religion, Nature and Culture* 13 (2): 181–207. doi:10.1558/jsrnc.36737.

Eck, Diana L. 2012. *India: A sacred geography*. Harmony Books. https://archive.org/details/indiaasacredgeography/page/n141/mode/2up.

Fisher, Karen. 2022. "Decolonizing rivers in Aotearoa." In *Indigenous women's voices: 20 years on from Linda Tuhiwai Smith's decolonizing methodologies*, edited by Tebrakunna Country, Emma Lee and Jennifer Evans. Zed Books.

Forster, Margaret. 2019. "He Tātai Whenua: Environmental genealogies." *Genealogy* 3 (3): 42. doi:10.3390/genealogy3030042.

Gobernación del Chocó. 2020. "Plan de desarrollo departamental 2020–2023 Chocó: Generando confianza." https://rap-pacifico.gov.co/wp-content/uploads/2020/07/choco-final-pddd.pdf.

Hastrup, Kirsten. 2009. "Waterworlds: Framing the question of social resilience." In *The question of resilience: Social responses to climate change*, edited by Kirsten Hastrup. Royal Danish Academy of Science and Letters.

Kramm, Matthias. 2020. "When a river becomes a person." *Journal of Human Development and Capabilities* 21 (4): 307–319. doi:10.1080/19452829.2020.1801610.

Krause, Franz. 2017. "Towards an amphibious anthropology of delta life." *Human Ecology: An Interdisciplinary Journal* 45 (3): 403–408. doi:10.1007/s10745-017-9902-9.

Krause, Franz, and Veronica Strang. 2016. "Thinking relationships through water." *Society & Natural Resources* 29 (6): 633–638. doi:10.1080/08941920.2016.1151714.

Linton, Jamie, and Jessic Budds. 2014. "The hydrosocial cycle: Defining and mobilizing a relational-dialectic approach to water." *Geoforum* 57: 170.

Macpherson, Elizabeth. 2023. "Can Western water law become more 'relational'? A survey of comparative laws affecting water across Australasia and the Americas." *Journal of the Royal Society of New Zealand* 53 (3): 395–424.
Mallabarman, Adwaita 1993 [1956]. *A river called Titash*. Translated by Kalpana Berdhan. University of California Press.
Mawdsley, Emma. 2005. "The abuse of religion and ecology: The Visha Hindu Parishad and Tehri Dam." *Worldviews: Global Religions, Culture, and Ecology* 9 (1): 1–24. doi:10.1163/1568535053628427.
McNeish, John A., and Jenny Moreno Socha. 2024. "Muddy waters: On the problematic political ecology of the Atrato ruling, Colombia." *The International Journal of Human Rights*: 1–21. doi:10.1080/13642987.2024.2324121.
Muller, Samantha, Steve Hemming and Daryle Rigney. 2019. "Indigenous sovereignties: Relational ontologies and environmental management." *Geographical Research* 57: 399–410. doi:10.1111/1745-5871.12362.
Nielsen, Kenneth Bo, Siddharth Sareen and Patrik Oskarsson. 2020. "The politics of caste in India's new land wars." *Journal of Contemporary Asia* 50 (5): 684–695. doi:10.1080/00472336.2020.1728780.
O'Donnell, Erin, Cristy Clark and Rachel Killean. 2024. "Rights and relationality: A review of the role of law in the human/water relationship." *Water Alternatives* 17 (2): 207–238.
Prasad, Indulata. 2022. "Towards Dalit ecologies." *Environment and Society: Advances in Research* 13: 98–120. doi:10.3167/ares.2022.130107.
Ranjan, Rahul. 2024. "Sadhu activism: Rights of rivers and facets of religious environmentalism in the Himalayas." *The International Journal of Human Rights*: 1–21. doi:10.1080/13642987.2024.2420244.
Salmond, Anne. 2014. "Tears of Rangi: Water, power, and people in New Zealand." *HAU: Journal of Ethnographic Theory* 4 (3): 285–309.
Siddalingaiah. 2003. *Ooru Keri*. Translated by S. R. Ramakrishna. Sahitya Akademi.
Sharma, Mukul. 2019. "'My world is a different world': Caste and Dalit eco-literary traditions." *South Asia: Journal of South Asian Studies* 42 (6): 1013–1030. doi:10.1080/00856401.2019.1667057.
Sharma, Mukul. 2022. "Caste, environment justice, and intersectionality of Dalit–Black ecologies." *Environment and Society: Advances in Research* 13: 78–97. doi:10.3167/ares.2022.130106.
Stats NZ. 2023. *2023 census*. https://www.stats.govt.nz/2023-census/#data.
Strang, Veronica. 2020. "Chapter 8: The rights of the river: Water, culture and ecological justice." In *Conservation*, edited by H. Kopnina and H. Washington. doi:10.1007/978-3-030-13905-6_8.
Tanasescu, Mihnea. 2020. "Rights of nature, legal personality, and Indigenous Philosophies." *Transnational Environmental Law* 9 (3): 429–453. doi:10.1017/S2047102520000217.
Tanasescu, Mihnea. 2022. *Understanding the rights of nature: A critical introduction*. Transcript Verlag.
Te Pou Tupua. 2023. "Tupua Te Kawa." https://www.tepoutupua.nz/tupuatekawa.
Tinirau, Rāwiri, Cherryl Smith, Meri Haami and Helena Rattray-Te Mana. 2020. *Te Oranga O Te Awa Tupua: A report prepared for Ngā Tangata Tiaki O Whanganui Trust*. Te Atawhai o Te Ao Charitable Trust.
Upadhyay, Kavita, and Bibhu Prasad Nayak. 2024. "Tracing the legal journey of petitions in the Uttarakhand High Court that became springboards for rights of

rivers and nature in India." *The International Journal of Human Rights*: 1–20. doi:10.1080/13642987.2024.2380850.

Waitangi Tribunal. 1999. *The Whanganui River Report*. GP Publications.

Whanganui Iwi and The Crown. 2014. *Ruruku Whakatupua – Te Mana O Te Iwi O Whanganui: Whanganui River Deed of Settlement*.

Wilson, Che. 2010. *Nga Hau O Tua, Nga Ia O Uta, Nga Rere O Tai* (Edition One), edited by H. Rainforth (H. Te Puna). Matauranga o Whanganui.

6 Legal innovations for rivers

Catalina Vallejo Piedrahíta, Nataly Montoya-Restrepo, Camilo Arango Duque and Julián Gaviria Mira

Introduction

The recognition of the rights of nature is an emerging legal concept that challenges traditional rights frameworks, which have historically centred on individual human interests. This shift raises critical questions about the legal, philosophical and practical implications of extending rights beyond humans. Some proponents argue that recognizing nature as a rights holder addresses the limitations of existing environmental protections and acknowledges the interdependence between humans and the natural world. Others, however, caution that such an expansion risks undermining the coherence of legal systems and diluting the meaning of rights, a concern raised by Marta Cartabia (2010), who warns that the proliferation of rights can lead to their political instrumentalization. Additionally, the concept challenges long-standing legal principles, such as the nature of legal personhood and the mechanisms for representing non-human entities in legal disputes. As Will Kymlicka (2024) argues, this debate ultimately demands a rethinking of justice – whether justice should remain a human-centred construct or be reimagined to account for broader ecological relationships: "One task facing defenders of more-than-human rights is to change public attitudes toward animals and nature, so that people come to understand and appreciate the value, significance, potentialities, and needs of the more-than-human world" (2024, 55). This shift not only raises theoretical questions but also requires a transformation in public attitudes. As discussions continue, it remains uncertain whether this paradigm shift offers a meaningful advancement in environmental governance or merely complicates rather than strengthens legal protections.

Since 2006, nearly 500 initiatives worldwide have sought to recognize the rights of nature, through constitutional reforms, legislation and judicial decisions (Rodríguez-Garavito 2024). Given the diversity of these approaches, it is important to explore the procedural and decision-making differences across regions to understand how these initiatives impact democratic debates and societal transformation. Traditional legal systems often privilege decisions made through legislative bodies, and scepticism exists regarding innovative judicial or administrative rulings, which are sometimes seen as counter-

DOI: 10.4324/9781003581208-6

majoritarian (Cruz Hermida et al. 2021, Salazar Ugarte 2011). However, the experiences of India, Aotearoa New Zealand and Colombia illustrate that the quality of democratic debate cannot be solely assessed through the procedure or authority behind decisions; the practical impact on environmental crises and social justice is paramount.

The recognition of nature's rights introduces a complex array of perspectives, ranging from those who view it as a path to tangible societal change, to others who warn of its potential to become a tool for political manipulation or a superficial response to deeper structural problems. Proponents see it as a transformative opportunity to address pressing environmental crises and expand notions of justice, while critics caution that it risks oversimplifying or depoliticizing ongoing struggles related to inequality, environmental degradation and resource extraction. As legal norms evolve, understanding how they interact with local ecosystems, cultural values and historical contexts becomes crucial in assessing their potential for meaningful impact. Ultimately, the question remains whether the recognition of nature's rights can lead to substantive change or whether it risks becoming another "ornamental" tool in the ongoing debate over environmental and social justice (González-Serrano 2024).

This chapter is structured to offer an exploration of legal debates on the emerging recognition of nature's rights, with a specific focus on rivers as new legal subjects. The first section, "Rivers as new legal subjects" delves into the arguments in favour of and critiques against affording legal personality to nature; it critically examines various legal models that have been proposed or implemented for recognizing ecosystems as rights-holders. The following section, "Why rights and why for nature?", explores the foundations of rights, questioning why rights are granted to nature and the philosophical underpinnings of these developments. The chapter then moves into a discussion of specific case studies, including Aotearoa New Zealand, India and Colombia, which highlight the application of the "legal personhood model" for three rivers. Through these cases, we explore the varying ways this model has been applied, focusing on the specific attributes granted under legal personhood, stewardship mechanisms chosen for representation, the processes through which rights are recognized and the rights and corresponding obligations conferred. A comparative analysis of these legal frameworks follows, offering an examination of the differing approaches and their implications. The chapter concludes with reflections on the potential and challenges of recognizing nature's rights, drawing insights from both theoretical debates and practical experiences across these three different jurisdictions.

Rivers as new legal subjects

The emergence of legal personality for all or parts of nature reflects a response to the limitations of existing laws and governance structures in addressing the exploitation and degradation of ecosystems (Macpherson et al. 2021, Kahui et al. 2024). These frameworks have arisen in regions where urbanization, agriculture, the extraction of natural resources and industrial activities have placed significant pressure on the environment (Kahui et al. 2024). Rights of nature frameworks vary widely in their design, legal structures and the roles of appointed representatives who can speak for ecosystems, ranging from public guardianship to granting ecosystems forms of legal personhood (Kahui et al. 2024, Kauffman and Martin 2021, Macpherson et al. 2021, Martin and Kauffman 2019).

The rights of nature call for a re-examination of the traditional boundaries that have historically separated nature from society and humans from non-human entities, to reshape public discussions and agendas (Rodríguez-Garavito 2024). For legal thought and practice, this new paradigm poses significant challenges, as it calls for the deconstruction of deeply ingrained liberal institutions and categories. However, the rights of nature also offers an opportunity to revitalize the democratic debate, update discussions on division of powers, and reassess the roles of judges, lawmakers and administrative bodies in environmental protection. Additionally, rights of nature is discussed by some authors as having potential to promote emancipatory uses of the law in social movements and foster interdisciplinary dialogue that incorporates alternative worldviews (Cartabia 2010, Rodríguez-Garavito 2024).

Rights of nature frameworks fundamentally transform the legal landscape by enabling legal action on behalf of ecosystems themselves, rather than solely on behalf of individuals whose health or property are harmed by environmental degradation (Stone 1972). This approach seeks to redefine the human-nature relationship, emphasizing ecosystems' intrinsic value and their own health or balance as independent legal concerns, rather than focusing exclusively on human impacts (O'Donnell 2020). By extending legal personhood to ecosystems, proponents seek to grant legal *agency* to natural entities historically treated as property – in domestic law – and as the territory over which states exercise sovereignty – in international law – (Natarajan and Dehm 2019). This shift enables ecosystems to engage in legal relationships independently (or interdependently) of human stakeholders, challenging traditional governance systems and emphasizing nature's intrinsic value.

Despite the promise of these frameworks, criticisms highlight potential limitations. For example, recognizing rights to ecosystems may anthropomorphize nature or perpetuate the human-nature dichotomy that characterizes modernity. Critics also question whether such recognition could inadvertently frame ecosystems as legally responsible entities, capable of incurring obligations or debts, similar to other legal persons (Martin and Kauffman 2019, Kauffman and Martin 2021).[1] Rivers, for instance, might be viewed as "helpless children" in need of protective guardians or as competitors for scarce resources alongside human agents (O'Donnell 2020).[2] Others argue that instead of generating a significant change in resource extraction and unsustainable practices, rights of nature often become framed within a discourse that, while symbolically powerful, fails to address the root causes of the socio-environmental conflicts faced by local communities, such as the political economy of extractivism (González-Serrano 2024, McNeish and Moreno Socha 2024, Melo-Ascencio 2024).

Models for the recognition of legal personality for nature vary across cases. Scholars like Kauffman and Martin distinguish between the "Legal Personhood Model" and the "Nature's Rights Model". In the latter, rights are granted broadly to nature as a whole, including the rights to exist, maintain ecological cycles and be restored when damaged. While this model allows anyone to advocate for nature, it does not obligate anyone to do so, making it fundamentally reactive (Kauffman and Martin 2021).

In contrast, the "legal personhood model" grants rights to specific ecosystems, such as rivers or forests, while also attributing characteristics typical of legal personhood. Under this model, nature not only holds rights but may also assume obligations, own property and participate in legal processes. Guardians are appointed to advocate on behalf of these ecosystems, ensuring proactive representation (Kauffman and Martin 2021).

While the "legal personhood model" offers a structured approach, its application varies across cases. This chapter examines these variations in the Atrato, Whanganui, Ganga and Yamuna rivers, focusing on (1) the specific attributes granted under legal personhood, (2) the stewardship mechanisms established for representation, (3) the processes by which these rights are recognized and (4) the rights and corresponding obligations conferred on ecosystems.

When ecosystems are granted legal status, they are recognized as independent entities with their own agency, similar to other non-human legal entities (Kahui et al. 2024). These include charitable organizations, corporations, trusts, labour unions, non-profit organizations, municipalities, local governments, states and international organizations. The legal fiction of such entities as "persons" allows them to be recognized by law as separate from the individuals or groups that make them up, enabling the legal fiction to own property, enter contracts, participate in legal actions or have intrinsic rights and interests. The fiction of the non-human legal person has its roots in Roman law. Over time, particularly in Western legal systems, this concept expanded to include a wide range of non-human entities, providing them with the ability to sue, be sued and engage in legal relationships independently of their individual members. This legal fiction has played a crucial role in shaping modern commercial and organizational law (Kurki 2019).

The roots of extending legal personality beyond the human person lay specifically in the classification made by the Roman jurist Gaius, who divided the law into three categories: persons (*personae*), things (*res*) and actions (*actiones*). While the Romans did not approach this distinction in a systematic way, it laid the foundation for the modern understanding of "legal personhood". Gaius's work, *Institutes*, written around 160 CE, was later incorporated into the *Corpus Iuris Civilis*, which became the basis for much of Western legal scholarship (Kurki 2019, 31) and the regulation of legal *persons* in continental civil law.

Kurki explains that the modern notion of legal personhood has evolved significantly over time. In Roman law, the term *persona* originally referred to the masks worn by actors in Greek and Roman theatre, symbolizing roles or qualities of human beings and thus referring to role or status within a hierarchical society. The concept of *caput* (literally meaning "head" and used to mean "legal standing") was the closest approximation to modern "legal personhood", but not all human beings were considered legal persons – slaves, for instance, were indeed human beings, but their legal personhood was neglected; they lacked *caput* and were treated as property rather than rights-holders. *Universitas* was initially the term used for associations and corporations, instead of *persona*. During the Middle Ages, corporations such as guilds and monasteries were granted legal recognition as *persona ficta*, allowing them to own property and enter contracts independently of individual members. The early modern period saw the emergence of a more systematic understanding of legal personhood, with scholars such as Hugues Doneau and Hermann Vultejus distinguishing between human beings (*homo*) and legal persons (*persona*), defining the latter as *entities with legal standing*. By the 19th century, authors like Savigny linked legal personhood to the ability to hold rights, reinforcing the idea that legal persons are the primary subjects of law (Kurki 2019).

Rooted in Roman law, legal personhood has traditionally been understood in two ways: (1) the capacity to hold rights and (2) the capacity to own property, with distinct interpretations in Continental and English legal traditions. In England, university colleges historically had the capacity to own property despite the absence of a fully developed doctrine of legal personhood. This was because English law, even before formalizing legal personality, recognized certain institutions – such as Cambridge colleges – as having independent legal existence for specific purposes, including property ownership. Unlike Continental law, which developed a systematic framework for legal personhood early on, the common law took a more pragmatic approach, granting entities legal capacities without categorizing them as legal persons in an abstract sense. It was not until the 19th century, under the influence of German jurisprudence, that English legal scholars began to systematize the concept of legal personality. Prior to this, entities like universities and municipal corporations were seen as having functional legal capacities rather than being full legal persons. This shift marked a move from a case-by-case approach to a more structured theory of legal personality, aligning English legal thought more closely with Continental traditions (Kurki 2019).

The recognition of nature as a subject of rights has gained increasing attention in contemporary legal and philosophical debates, particularly in response to the growing environmental crisis. One central question that arises in this context is *Why should nature have rights?* This question can be approached from various perspectives, including normative, empirical and analytical frameworks. The next section delves into the philosophical and legal justifications for the recognition of rights in general, and for nature in particular, exploring the implications of such recognition.

Why rights and why for nature?

Rights are fundamental to legal, political and moral discourse, commonly understood as claims or entitlements held by individuals, groups or entities with legal personhood, which impose corresponding duties on others (Raz 2006). This section examines what legal rights, once recognized by a political community, can achieve and explores the justifications for their recognition.

Why rights?

Beyond the reasons that justify protecting a particular interest or entity through the granting of rights – a question to which we will return later – it is important to first clarify what rights are and what it means to assign rights to nature. To address this issue, the framework of legal positions proposed over a century ago by Wesley Newcomb Hohfeld proves particularly useful. Hohfeld provided a clear and effective explanation of what a right is – or, more precisely, what is meant when one says that someone or something has a right. According to Hohfeld's scheme, when legal discourse affirms that an entity has a right, this statement implies that the entity possesses either a liberty, a claim, a power or an immunity (Hohfeld 1917).

In summary, and following Carl Wellman's reconstruction, a liberty is simply "the absence of any contrary legal duty. Thus, person P has a legal liberty to do act A if and only if P has no legal duty not to do A." A claim is "the logical correlative of a corresponding legal duty of some second party. Person P has a legal claim against person Q that Q do act A if and only if Q has a legal duty to P to do act A." A power is "the ability to bring about some change in the legal situation by performing some operative action. Thus, P has the legal power to effect some legal consequence C if and only if, were P to perform operative act A, then C would be in effect". Finally, an immunity refers to the inability of others to exercise power (as just defined) over the rights-holder (Wellman 1995, 12–37).

Hohfeld's framework clarifies what having a right entails or enables. The question that follows is whether, when advocating for the recognition of rights for nature, we are referring to conferring liberties, claims, powers and immunities upon it, or whether we are speaking of only some of these legal positions. While this is not a closed discussion, the cases analysed in the final section of this chapter illustrate that all the legal positions described by Hohfeld are at play when discussing the rights of nature, though this requires some preliminary clarification.

Granting rights to nature entails multiple legal and conceptual implications. These legal changes aim to protect ecosystems such as the Atrato River, the Whanganui River and the Ganges and Yamuna rivers. However, recognizing rights for these ecosystems goes beyond merely establishing that humans have certain duties towards their preservation – meaning that, in Hohfeldian terms, these ecosystems have a claim to protection, which corresponds to our duty of care. The recognition of nature's rights has also served as a tool for leveraging social groups' power, such as the Māori people in Aotearoa New Zealand and the Afro-descendant and Indigenous riverine communities in the Chocó biogeographic region of Colombia. This implies that nature's rights also function as powers, in the Hohfeldian sense, enabling these communities to participate in decision-making processes that affect the river basins.

The cases of the Whanganui River in Aotearoa New Zealand and the Ganges and Yamuna rivers in India show that recognizing rights for certain ecosystems entails not only claims and powers but also "the duties and liabilities of a legal person", as stated in Article 14 of the Te Awa Tupua (Whanganui River Claims Settlement) Act. This aspect has been subject to significant criticism (see Kauffman and Martin 2021). The recognition of legal personhood for rivers raises concerns about whether such entities should bear legal responsibilities in the same way as human or corporate legal persons, illustrating the complexity of this legal framework and its implications.

The relationship between rights and duties is complex, with different theories offering varying perspectives. Some see rights as always tied to corresponding duties, while others argue that this connection is not always fixed or straightforward. Rights often serve as grounds for duties, but they do not always create a direct one-to-one relationship (Raz 2006). For example, a right may justify imposing duties on others, but the specific nature of those duties can change depending on the social and legal context. Take the right to political participation – while it may lead to duties, such as a government's obligation to disclose information, this duty only exists within certain institutional frameworks. Rights are therefore dynamic, shaping and evolving duties in ways that align with the interests they protect, rather than always having an exact counterpart (Gewirth 1986, Raz 2006).

Hohfeld's framework clarifies what rights can allow, but a deeper question remains: why do we recognize certain rights in the first place? Three main theories offer different answers – status-based, instrumental and contractual. Status-based theories, rooted in natural rights traditions, argue that rights come from fundamental human qualities like autonomy and dignity. They treat rights as strong and nearly absolute, though they can be difficult to adapt to different contexts. Instrumental theories take a more practical approach, viewing rights as tools for achieving desirable social outcomes, such as fairness or wellbeing. However, this flexibility can make rights seem uncertain, as their justification depends on changing circumstances. Contractual theories suggest that rights are based on principles that rational individuals would agree to under fair conditions, aiming to balance individual freedoms with broader societal needs (Herstein 2023).

One of the trickiest aspects of expanding rights to a broader range of legal subjects is the tensions between them. Conflicts of rights occur when rights are incompatible, either due to conflicting claims or conditions for their application (Wellman 1995). The distinction between "abstract rights" (broad principles) and "concrete rights" (specific applications) helps explain this: while abstract rights can coexist, conflicts arise in concrete cases, like competing claims to possess the same item. To resolve conflicts, rights may be redefined to avoid overlap, or a hierarchy of rights can be established by courts, with some rights taking precedence over others. When no hierarchy exists, courts may balance the interests behind each right or determine which best promotes those interests (Thomson 1990, Waldron 1993 in Herstein 2023).

Why for nature?

A key question in recognizing nature as a subject of rights is: why should nature have rights at all? For example, if a legal system says "the river has a right to conservation", this could mean at least three things: a) the river is valuable to humans thus should be protected; b) the river is protected by a law that grants it the right to stay in a natural state; or c) the river has a moral right to remain in a certain state.

Option a) is centred on the rights and interests of humans. It serves a utilitarian function, meaning the river only has rights because it benefits humans. In this view, human needs would take priority over the balance of the ecosystem, and parts of the river that do not directly benefit humans could be ignored. However, it does not make sense to grant rights to nature just to serve human interests. If that were the case, it would be more logical to focus on human rights instead. Option b) suggests a legal construction where the river's right is justified either by its usefulness to humans a) or its moral worth c). If justified by a), the issues of anthropocentrism and utilitarianism still apply.

Option c) is more philosophical and argues that the river has an inherent right to conservation simply because it exists. The intrinsic or inherent value of the river raises some challenges. Does not valuation require a valuing subject and a subject valued? For example, would viruses or bacteria that human societies combat daily also have inherent value and hold rights based on their existence? What is the foundation for recognizing legal rights to some existing entities and not to others? These are complex questions that require careful consideration.

To escape the trap of granting rights to nature just to serve human interests, a key philosophical justification for recognizing the rights of rivers may lie in attributing *ecosystemic value* to the river. Once it is understood that human communities exist within ecosystems and are *part of them*, societies – aware of this interconnection and the often-broken relational bonds between nature and human life – begin to speak on behalf of the ecosystem, not just their own species as separate. Furthermore, there are other ways of valuing nature or rivers that do not rely on utilitarian reasoning, such as aesthetic, religious or poetic valuations. The utilitarian approach often reflects the human-nature separation. In contrast, poetic valuation, feeling, as Barbaras (2019) points out, helps bridge this gap, to partially return from exile, reconnecting the human being with *physis*. Aesthetic experiences allow us to engage with the object in a different way and open ourselves to it. This is why the worldviews of Indigenous and other rural groups and vernacular communities have been pivotal in granting rights to nature. In many cases, their relationship with nature has embodied this holistic character.

The recognition of nature's rights is often seen as requiring human representation, meaning that humans would determine what these rights entail and act on behalf of nature in their defence. However, simply granting rights to nature – such as to rivers or forests – does not automatically ensure stronger protection. If the underlying metaphysical framework remains unchanged, and a utilitarian perspective continues to dominate, the recognition of nature's rights could become more formal than substantive, with little practical effect. In this view, nature would have rights, but these rights could mirror a limited version of individual human rights, without addressing the inherent value of other life forms and the deeper relational aspects between humans and ecosystems. As Ximena González-Serrano (2024) puts it, this could amount to little more than "ornamental legal frameworks".

On the other hand, some argue that incorporating more holistic or poetic, religious and aesthetic perspectives of nature could foster more meaningful changes in environmental protection. This might involve either granting rights to nature itself or expanding the scope of human rights to encompass broader ecological concerns. In this sense, the specific approach – whether by recognizing nature's rights or broadening human rights – might be less important than ensuring that the theoretical foundation is robust and aligned with the broader goals of environmental protection.

In addition, the recognition of nature's rights requires them to be integrated into coherent constitutional and legal systems. This process of incorporating such rights into existing legal frameworks may require adjustments, making it necessary to explore other institutional approaches to ensure the effective implementation of these rights. The aim is to find a way to introduce these new rights while avoiding potential legal imbalances or resistance. This raises the question of which category of rights is most appropriate for nature. One perspective is that the rights of nature could be understood as *collective rights*. [3] While biology can identify and categorize individual species within an ecosystem (Wilson 2021), the concept of an "individual" in legal philosophy typically refers to the capacity for self-awareness, responsibility and autonomy. Therefore, traditional categories such as fundamental, personal, or subjective rights – designed to protect individual human subjects – may not be suitable for nature.

In contrast, the concept of collective rights could be more compatible with the nature of ecosystems, which are inherently collective rather than individual entities. If nature is to be considered a subject of rights, it may be more appropriate to think of it as a collective entity that is represented by humans who belong to the ecosystem. Human beings who are part of that community of life within an ecosystem could then act as representatives, going to court on behalf of the collective. In this context, the concept of a "community of rights" or a "collective of rights" might better reflect the nature of these legal claims, rather than framing it as a "subject of rights".

This idea brings us closer to existing collective rights, such as *the right to a healthy environment*. An interesting way to nuance human-centred approaches that exclude ecological relations within this context is to consider that when the right to a healthy environment is enshrined, the adjective "healthy" should not be predicated solely on the health of the human community but on the health of the environment itself. In other words, if the right to a healthy environment were understood as the right of a community to live in a *healthy ecosystem* – not just healthy for humans but healthy in and of itself – then we would be closer to considering an ecosystem's right to conservation.

It is important to note that health, as traditionally defined in legal systems, has been a subjective criterion primarily related to human wellbeing. However, a healthy environment for humans does not necessarily equate to a healthy ecosystem. For example, areas rich in biodiversity may thrive as healthy ecosystems, but non-native human populations could quickly fall ill in such environments. Conversely, heavily degraded ecosystems might still support human communities with relatively high public health standards. Consider, for instance, that most freshwater fish could not survive in our drinking water tanks; chlorine, in amounts harmless to humans, would be lethal to them (Zillich 1972). If this ecocentric interpretation of the right to a healthy environment were adopted, it could also be framed as a human right, which would open the possibility of applying legal mechanisms for its protection. This could involve judicial bodies such as the Inter-American Court of Human Rights, the African Court and the European Court of Human Rights,[4] as well as specific national legal systems for the protection of fundamental rights, such as writs of protection or constitutional protection actions, with their inherent urgency and priority.

In conclusion, the question of why nature could or should have rights is complex and involves multiple perspectives, encompassing philosophical, legal and practical considerations. Some approaches, grounded in utilitarianism, link rights primarily to human interests, while others argue for recognizing nature's intrinsic value, informed by cultural, aesthetic or ecological views. The recognition of nature's rights requires a robust philosophical foundation to prevent superficial gestures that may limit the potential for meaningful ecological change. Furthermore, nature's rights might be better understood not as individual rights but as collective rights, reflecting the interconnectedness of ecosystems and the communities they support. Framing nature's rights in this way aligns with existing legal categories, such as the collective right to a healthy environment and could enable broader legal recognition and protection. This shift could promote a more ecologically oriented legal framework, extending beyond human-centred norms and aiming to protect ecosystems for their inherent value, which would benefit both nature and human individuals and communities.

Building on this discussion of nature's rights, legal pluralism offers a valuable lens for understanding how different legal traditions and normative frameworks interact in recognizing and enforcing environmental protections. Many legal systems already incorporate pluralistic elements, where Indigenous, customary and state legal orders coexist and influence one another. This is particularly relevant in the context of nature's rights, as several non-Western traditions – such as Māori and also Andean legal frameworks in Ecuador and Bolivia – have played a pivotal role in shaping contemporary legal debates on the subject. Furthermore, the recognition of nature's rights has been shaped by legal transplants and transnational legal interactions. The recognition of rivers as legal persons, for example, has travelled across jurisdictions, from Aotearoa New Zealand to India and Colombia, illustrating how legal ideas are not simply transferred but rather adapted to the specific legal, political and cultural contexts in which they are received. This highlights the dynamic nature of transnational legal processes in a globalized world, where legal pluralism provides a crucial framework for understanding both the diffusion of legal norms and the localized negotiations that shape their implementation. Examining legal pluralism allows for a deeper exploration of how diverse legal traditions engage with environmental governance, how conflicts between different legal orders are dealt with, and how pluralistic legal approaches might enhance the implementation of nature's rights within existing legal structures.

Legal pluralism and the recognition of nature's rights

Legal pluralism refers to the coexistence of multiple legal systems within a single social or political space, challenging the traditional notion that law is solely the product of the state. This concept recognizes that various normative orders – such as religious laws, customary practices and transnational legal frameworks – operate alongside or even in competition with state law (Twining 2021). Boaventura de Sousa Santos emphasizes that modern societies are shaped by a plurality of legal orders, which are interrelated and socially distributed in different ways. This perspective is particularly relevant in postcolonial and multicultural societies, where multiple legal traditions persist despite the dominance of Western state law (de Sousa Santos 1995). In the case studies we examine, interactions between Indigenous and Black communities' legal systems and state law illustrate how legal pluralism operates within specific socio-legal contexts.

A central debate in legal pluralism concerns the extent to which non-state legal systems should be recognized and integrated within formal legal structures. Rachel Sieder (2019) highlights that, for example, in Latin America, the recognition of Indigenous and customary law has led to complex interactions between state and non-state legal systems, raising questions about legitimacy, authority and the protection of human rights within these plural legal contexts. This discourse also engages with the concept of *interlegality*, introduced by de Sousa Santos (1995), which describes the intricate ways in which multiple legal and normative orders interact and influence one another.

Legal pluralism has also shaped debates on *legal diffusion* and *transplantation* of norms from one place to another. Foreign legal norms are never introduced into an empty legal space but must inevitably interact with pre-existing local arrangements, often deeply embedded in institutionalized normative orders (Twining 2006). The case of riverine rights exemplifies this dynamic, as legal frameworks recognizing rivers as subjects of rights have been influenced by earlier developments in Aotearoa New Zealand yet have undergone a process of adaptation to Colombia's and India's constitutional and legal traditions. Legal transfers do not occur in isolation but instead adapt to the specific legal contexts that receive them, illustrating how legal pluralism functions within a transnational understanding of law in a globalized world (Michaels 2009). Twining (2012) expands on this view, asserting that legal pluralism broadens the understanding of law beyond state-centric perspectives by acknowledging the role of non-state actors and informal legal mechanisms in shaping societal norms and dispute resolution processes.

Scholars also debate whether legal pluralism should be understood as a strong or weak phenomenon. Twining (2021) explains that some perspectives advocate a strong version of legal pluralism, which sees legal orders as independent and, at times, conflicting, while others support a weaker version, where different legal systems interact within a hierarchical framework led by state law. Another key discussion concerns the legitimacy and authority of non-state legal systems, particularly when customary or religious laws conflict with human rights principles or democratic values. This raises critical questions about how to balance respect for cultural diversity with the protection of fundamental rights and the risks of reinforcing power imbalances within certain communities (Twining 2021).

Additionally, legal pluralism has sparked broader discussions on governance and legal theory. Some scholars view it as a means of improving access to justice, particularly in postcolonial and Indigenous contexts, where state law may be inaccessible or ineffective. Others caution that legal pluralism can lead to legal fragmentation and inconsistencies, making it difficult to establish which authority decides a matter, what the applicable norm is, how to decide when there is a conflict of norms or how to ensure legal certainty and equality before the law. The complexities of legal pluralism are further exacerbated by globalization, as transnational legal regimes – such as international human rights law, trade law and environmental law – interact with national and local legal orders, creating intricate legal landscapes that challenge traditional notions of sovereignty and legal authority (Twining 2012, 2006, 2021).[5]

These debates underscore the significance of legal pluralism as both an analytical framework and a policy concern in an increasingly interconnected world. The next section explores the complexities of recognizing the rights of nature, drawing from the experiences of countries such as India, Aotearoa New Zealand and Colombia. It highlights the importance of context-specific legal frameworks, the involvement of diverse stakeholders and the need for tangible actions to address environmental crises. While the "legal personhood

Figure 6.1. Fisherman on the Atrato River. Photo by Daniel Alfonso León.

model" for nature provides a structured approach, its application varies significantly across different cases. The following section examines these variations by focusing on the attributes granted under legal personhood, the stewardship mechanisms for representation, the processes through which these rights are recognized and the rights and obligations conferred on ecosystems.

Rivers as legal subjects in Aotearoa New Zealand, Colombia and India

India, Aotearoa New Zealand and Colombia have each recognized rivers as legal subjects in some capacity, and all three countries experienced prior socio-legal mobilization efforts aimed at improving river governance. However, their approaches differ significantly, reflecting distinct legal traditions, cultural contexts and implementation challenges. India and Colombia's frameworks are shaped by public interest litigation and judicial activism, while Aotearoa New Zealand emphasizes Treaty-based governance and legislation that promotes shared decision-making with Indigenous communities. Colombia has integrated the concept of biocultural rights into its constitutional framework, establishing a combined commission of community and government river guardians. In contrast, India has assigned river guardianship exclusively to state representatives under the *parens patriae* doctrine. The following sections will analyse the specific legal mechanisms employed in each case and, in the final section, offer a comparative exploration of these approaches.

Aotearoa New Zealand

In Aotearoa New Zealand, or Aotearoa in Māori, the legal recognition of nature's personhood has developed through a unique interplay of historical, legal and cultural frameworks, particularly in relation to the Treaty of Waitangi. Signed in 1840 between the Crown and Māori chiefs, this foundational document established the basis for a bicultural nation. Unlike Colombia and India, Aotearoa New Zealand lacks a single written constitution; instead, its constitutional norms are dispersed across legislation, common law and customary practices. This fragmented structure complicates the recognition of environmental rights. For example, while the Aotearoa New Zealand Bill of Rights Act 1990 protects some human rights, it does not constitutionally enshrine environmental rights, such as the right to a healthy environment or the rights of nature (Macpherson et al. 2021).

Aotearoa New Zealand's approach to environmental governance has been shaped significantly by Treaty settlements, which acknowledge Māori ancestral relationships with natural resources and establish new governance structures. A major development in this evolution was the Te Awa Tupua Act 2017, which granted legal personhood to the Whanganui River (Te Awa Tupua). This Act formally recognizes the river as a legal entity with rights, powers, duties and liabilities. Te Pou Tupua is a body established to represent the river. It is made up of one member named by the Whanganui *iwi* and one by the Crown, requiring consensus in decision-making. The Act seeks to reflect Māori values by emphasizing the river's physical and spiritual interconnectedness.

However, the Te Awa Tupua model is a political compromise. While it acknowledges the river's legal personhood, the Crown retains ownership of the water, and the Act does not fundamentally alter existing governance frameworks, such as those under the Resource Management Act 1991 (Macpherson et al. 2021). Despite these limitations, the model represents a shift towards integrating Indigenous perspectives and legal pluralism into environmental governance (Cribb et al. 2024). Additionally, broader regulatory reforms, including updates to the National Policy Statement for Freshwater Management and ongoing revisions to the Resource Management Act, further emphasize Māori worldviews in environmental policy.

A key takeaway from New Zealand's experience with riverine rights is that small-scale or *ad hoc* developments, which embody gradual change and incremental progress, can lay the groundwork for more significant environmental reforms. In this context, Te Awa Tupua can be viewed as a constitutional innovation signalling potential pathways towards transformative change (Macpherson et al. 2021, 463).

Colombia

Colombia's 2016 Atrato River ruling (Constitutional Court of Colombia 2016) marks a shift in its constitutional jurisprudence, addressing socio-environmental issues by recognizing the river as a "subject of rights". Rooted in the civil law tradition and the 1991 Constitution, this decision integrates human rights, cultural identity and the natural world, establishing "biocultural rights" shared by local communities and the river (Cuesta Mosquera and Vallejo Piedrahíta 2024, González-Serrano 2024, Macpherson et al. 2021, Macpherson 2020, Wesche 2021).

Biocultural rights, as defined by the Constitutional Court, grant Indigenous and Afro-descendant communities the right to manage and protect their territories and natural resources according to their laws and customs. These rights acknowledge the inseparable link between nature, its resources and the culture of these communities (Constitutional Court of Colombia 2016).

The 1991 Colombian Constitution, developed during re-democratization, emphasizes social rights, multiculturalism and environmental protection. It is often called an "ecological Constitution" for its extensive environmental rights provisions, including Articles 79 and 80, which define the state's duty to conserve the environment and promote sustainable development (Bonilla Maldonado 2006). Key legislative measures, such as Law 70 of 1993, grant the rights of Afro-descendant communities to collective land ownership and self-governance (Gómez Erazo 2012, Congreso de la República de Colombia 1993, Mosquera Perea and Rentería Bonilla 2024, Pérez Caballero et al. 2024).

In the 2016 Atrato River case, the Constitutional Court recognized the river as a subject of rights in response to claims from Afro-descendant and Indigenous communities. These communities argued that their rights to life, health, livelihoods, water and a safe environment were violated due to the river's degradation caused by extensive illegal gold and sand extraction by transnational organized crime. The primary sources of river degradation identified included changes to the riverbed, embankment alterations and mercury pollution from gold mining (Constitutional Court of Colombia 2016).

The Court granted the river legal personality as a subject of rights, granting it four rights: to protection, conservation, maintenance and restoration and mandated the state and local communities to take action (Cantillo Ramírez et al. 2019, Castañeda Ruiz et al. 2019). Inspired by Aotearoa New Zealand's Te Awa Tupua framework (Cuesta Mosquera and Vallejo Piedrahíta 2024, Macpherson et al. 2021), the decision also established a guardianship commission and expert panel to foster community participation in environmental governance. This ruling set a precedent for recognizing other natural entities as legal subjects, including the Amazon, páramos and various rivers and ecosystems (Vallejo Piedrahíta et al. this volume, Sánchez Jaramillo 2022, Sánchez Zapata 2022).

Judicial decisions now cover ecosystems that span about 72% of Colombia's territory, meaning these areas must be considered subjects of rights, i.e., participants in their governance, not just subjects of administrative management (Sánchez Zapata 2022). However, implementation remains a challenge. Colombia's legal and procedural frameworks do not fully accommodate the rights of natural entities, which has led to an "implementation gap" that hampers enforcement. Additionally, the country's weak environmental institutions and ongoing armed conflict complicate the protection of both communities and ecosystems (Cuesta Mosquera and Vallejo Piedrahíta 2024, González-Serrano 2024, Sánchez Zapata 2022, Wesche 2021).

India

In India, the recognition of nature as a legal person arose from a combination of constitutional principles, environmental legislation, religious influences and judicial activism. India's legal framework, based on common law inherited from British colonial rule, was initially secular and did not prioritize environmental protection. However, over time, a series of constitutional amendments and judicial decisions paved the way for recognizing nature's legal subjectivity (Macpherson et al. 2021, Upadhyay and Nayak 2024).

The Indian Constitution, drafted in 1950, initially lacked provisions for environmental protection, but the 42nd Amendment in 1976 introduced Article 48A, which obligated the state to protect the environment and Article 51A(g), which required citizens to safeguard natural resources. Although these provisions were unenforceable, they highlighted the growing importance of environmental concerns. Additionally, laws like the Water Act of 1974 and the Environment Protection Act of 1986 laid the groundwork for public interest litigation, which enabled courts to hear cases on environmental harm even without direct personal injury. The expansion of Article 21, guaranteeing the right to life, played a key role in elevating environmental protection to a constitutional level. Landmark rulings in the 1990s, including a 1991 decision that explicitly included the right to pollution-free water and air within Article 21, further affirmed environmental rights as essential to human dignity (Macpherson et al. 2021). Public interest litigation, introduced in the 1970s, allowed citizens to advocate for environmental protection on behalf of marginalized communities. The Supreme Court's decisions, including cases concerning industrial pollution in the Ganga, solidified the principle that a clean environment is a fundamental human right (Upadhyay and Nayak 2024).

In 2017, the Uttarakhand High Court recognized the Ganga and Yamuna rivers as "living entities" with legal rights. This ruling drew on Hindu beliefs about the sanctity of rivers, and the court appointed guardians to act on behalf of the rivers, like the doctrine of *parens patriae*, where the state serves as a guardian for those unable to represent themselves. This legal personhood approach mirrored Christopher Stone's (1972) argument for granting legal standing to natural entities, such as trees and rivers, and suggested that the state or citizens could act on behalf of the environment when its interests are at risk.

While the ruling was a novel fusion of legal, cultural and religious principles, it faced challenges related to enforceability and jurisdiction, particularly as the Ganga flows through multiple subnational states and into Bangladesh. At the time of writing, the Supreme Court had stayed the ruling pending further review.

Comparative analysis

The legal recognition of rivers as persons in Aotearoa New Zealand, Colombia and India reflects distinct legal traditions and cultural frameworks, each contributing unique elements to the evolving discourse on the rights of nature. These cases illustrate different ways in which legal systems can integrate environmental protection, Indigenous and local community rights and state guardianship into their legal personhood models. While they present opportunities for further development, they also raise complex questions about implementation, governance and the extent to which legal personhood for nature can lead to substantive changes in environmental law and policy.

Colombia: Biocultural rights and the protection of rivers

Colombia's legal recognition of nature's rights is deeply intertwined with the concept of biocultural rights, which frames environmental protection as inseparable from the rights of riverine communities. The 2016 ruling on the Atrato River marked a turning point in two ways: by recognizing the river as a subject of rights, the state is required to ensure its protection, restoration, maintenance and conservation. And furthermore, by introducing the concept of biocultural rights, this approach acknowledges that environmental harm disproportionately affects historically marginalized communities and positions legal personhood as a tool for environmental justice. The Court's decision goes beyond a purely ecological rationale, recognizing the cultural, spiritual and economic interdependence between local communities and their ecosystems.

To enforce this ruling, the Court mandated the creation of a commission of guardians and an expert panel, ensuring that both state institutions and local communities play a role in river governance. This decision set a precedent for the legal recognition of other natural entities in Colombia, including the Amazon rainforest, other rivers and high-altitude wetlands (*páramos*). However, despite this expansive vision, significant challenges remain in implementing biocultural rights. Colombia's legal and institutional frameworks are not yet fully equipped to enforce the interdependent rights of natural entities and people, state institutions continue to be omissive and extractive industries continue to exert pressure on ecosystems. The persistence of armed conflict in regions like the Atrato Basin further complicates efforts to establish effective collaborative environmental governance (Vallejo Piedrahíta et al. this volume).

The Colombian case raises critical questions about whether granting legal personality to nature is sufficient without broader structural reforms, including stronger environmental institutions and increased local governance capacity. Future developments may explore ways to better integrate biocultural rights into national environmental policies, ensuring that legal recognition translates into practical enforcement and community empowerment. As environmental degradation and social inequalities persist, Colombia's experience provides an important perspective on how the rights of nature can be conceptualized within a broader social justice framework.

India: The best interests of nature and legal guardianship

India's approach to river personhood takes a different legal path, drawing on the doctrine of *parens patriae*, which allows the state to act as a guardian for entities unable to represent themselves. The 2017 Uttarakhand High Court ruling granted legal personhood to the Ganga and Yamuna rivers, citing both environmental concerns and the rivers' religious significance for Hindus. The court's reasoning drew on Hindu traditions, likening the legal status of the rivers to that of idols, which have long been recognized as legal persons in Indian law. By framing the rivers as vulnerable entities in need of protection, the ruling positioned the state as their steward, echoing legal protections used for children and persons with disabilities both in Indian and other legal systems.

This model aligns with legal protections in European and English law, where courts appoint guardians *ad litem* ("for the case") to represent the best interests of minors or individuals lacking decision-making capacity. It also draws from Christopher Stone's (1972) seminal argument in "Should trees have standing?", which proposed that natural entities could be granted standing in court, with legal representatives acting on their behalf. Echoing the "best interests of the child" legal principle, by applying a "best interests" framework to nature, this approach suggests that rivers, like vulnerable persons, should be protected from harm by legal advocates ensuring that their rights are *prioritized* against competing economic interests.

While this model presents an intriguing legal innovation, it also raises complex jurisdictional and governance issues. The Ganga and Yamuna rivers flow through multiple states and extend into Bangladesh, creating significant enforcement challenges and jurisdictional conflicts. The High Court's decision to stay the ruling reflects concerns about the enforceability of river personhood across multiple states and the challenge of reconciling religious symbolism with constitutional secularism. Furthermore, unlike Colombia's biocultural approach, which emphasizes community involvement, India's model centralizes authority within the state, raising questions about how legal guardianship structures should be designed to ensure effective and inclusive environmental governance and participation.

Future legal developments might explore alternative forms of guardianship, including the possibility of environmental ombudspersons or independent river councils, which could operate with greater autonomy from state political interests. Additionally, the application of legal personhood to transboundary rivers highlights the need for coordinated governance mechanisms that address enforcement challenges beyond national borders.

Aotearoa New Zealand: Integrating ancestral values into river governance

Aotearoa New Zealand's Te Awa Tupua Act (2017) represents a model of river personhood that explicitly integrates Indigenous governance and legal pluralism. The recognition of the Whanganui River as a legal person acknowledges Māori perspectives, which view the river as a living and indivisible entity. Unlike traditional legal frameworks that divide rivers into separate legal components – such as water rights, land ownership and resource management – this approach treats the river as a singular living being, encompassing both its physical and spiritual dimensions. The governance structure established under the Act requires co-management between the Whanganui tribes (*iwi*) and the Crown, creating a decision-making process based on consensus.

Despite its recognition of Indigenous values, this model remains a political compromise. The Crown retains ownership of water resources, and the Act does not fundamentally alter existing water governance frameworks, such as those established under the Resource Management Act 1991. This raises questions about whether legal personality is being used primarily as a symbolic acknowledgement of Māori worldviews or whether it has the potential to reshape water governance in a more substantive way. Cribb et al. (2024, see also Cribb this volume) argue for the latter by pointing out the new and inclusive governance institutions being developed as a direct effect of the Te Awa Tupua Act.

In our view, Te Awa Tupua signals an incremental but meaningful shift towards a legal system that balances ecological, political, cultural and legal considerations. Ongoing reforms in freshwater management policies indicate a growing shift towards incorporating Māori perspectives into environmental governance, suggesting that the legal personality framework could evolve towards stronger Indigenous governance rights.

Expanding the rights of nature framework

These three legal models – biocultural rights in Colombia, legal guardianship in India and Indigenous co-governance in Aotearoa New Zealand – offer different pathways for developing the rights of nature movement. The Colombian approach highlights the connection between nature's rights and social justice, proposing a model that could be expanded in other contexts where local communities play a key role in environmental stewardship in contexts of expanding extractivism. The Indian case demonstrates the potential for state-led guardianship mechanisms, raising important questions about how legal personhood can function in national legal systems. Meanwhile, Aotearoa New Zealand's model presents a framework for ecological legal pluralism, integrating Indigenous governance into state environmental law, suggesting possibilities for co-management structures that blend traditional and state legal systems.

While each of these approaches introduces valuable legal innovations, they also reveal ongoing debates about the effectiveness, enforceability and implications of granting legal personality to rivers. Future developments in rights of nature frameworks may draw on elements from each of these models, adapting them to local legal and ecological contexts to create more robust and enforceable protections for the natural world. Whether legal personhood ultimately leads to greater environmental protection or remains a symbolic legal recognition depends on how these frameworks continue to evolve and how legal systems address the challenges of recognition of rights, implementation, governance and enforcement.

Conclusion: Legal personhood for nature and its diverging models

The recognition of rights of nature frameworks reflects an evolving legal approach to environmental governance, seeking to address the limitations of traditional legal systems in protecting ecosystems. By granting legal personhood to rivers, these frameworks propose a shift from viewing nature as an object of human regulation and appropriation to acknowledging it as a legal subject capable of holding rights. This development draws from the historical evolution of legal personhood, a concept shaped by Roman law, Continental private law traditions and Anglophone jurisprudence, which has extended legal status to entities such as corporations, municipalities and religious institutions. While these approaches share the objective of integrating nature into legal frameworks, they diverge in their theoretical foundations, governance mechanisms and enforceability, raising questions about their practical implications and legal coherence. Moreover, these models must be understood through the lens of legal pluralism, as they emerge within distinct legal cultures, engage with multiple normative orders and undergo processes of legal transplantation and adaptation in different contexts.

The recognition of rivers as legal persons in Aotearoa New Zealand, India and Colombia illustrates different ways in which legal systems have engaged with this concept. Aotearoa New Zealand's Te Awa Tupua Act (2017) establishes a co-governance model in which the Whanganui River is recognized as a living and indivisible entity, integrating Māori legal values and state governance. *Te Pou Tupua*, a guardianship body representing both the Whanganui *iwi* (tribes) and the Crown, speaks and acts on behalf of the river. Unlike Colombia's Atrato River, which was declared a subject of rights but cannot own property or engage in transactions, the Whanganui River can, giving it a broader range of legal capacities. However, in Aotearoa New Zealand, the Crown retains ownership of water, and the Act does not introduce broader shifts in resource management laws. Colombia's Constitutional Court ruling on the Atrato River (2016) is framed within "biocultural rights", a legal approach that connects the recognition of nature's legal personhood to the rights of Indigenous and Afro-descendant communities. This model introduces a legal framework in which ecosystem protection is linked to the social

Table 6.1. Comparative analysis of nature's rights recognition in Aotearoa New Zealand, Colombia and India. Table by authors.

Country & River	*Legal Recognition*	*Stewardship & Governance*	*Legal Framework & Justification*	*Rights & Obligations*	*Challenges in Implementation*
Aotearoa New Zealand – Whanganui River	Recognized as a living whole and legal person with rights, powers, duties and liabilities under the Te Awa Tupua Act (2017).	*Te Pou Tupua* (two representatives: one from the Whanganui Iwi and one from the Crown) acts as the river's guardian, requiring consensus decision-making.	Rooted in the Treaty of Waitangi, acknowledging the river as an indivisible and living whole, reflecting Māori worldviews.	The river holds a legal status comparable to a person, but no specific rights were mentioned. The Act ensures that governance respects Indigenous perspectives.	Existing water governance structures and political compromises limit transformative potential in environmental management.
Colombia – Atrato River	Recognized as a "subject of rights" but not explicitly as a legal person (*persona jurídica*) (Constitutional Court Decision T-622/2016). The ruling does not grant the river the ability to sue, hold property or enter transactions.	Commission of River Guardians includes government agencies and local communities, with advisory experts (e.g., Humboldt Institute, WWF Colombia).	Emerged from a constitutional claim (public interest litigation) by Afro-descendant communities due to state inaction on environmental degradation and human rights violations.	The state, society and individuals have specific obligations to *protect, conserve, restore, maintain* and sustainably manage the river and surrounding ecosystems.	Weak institutional capacity, armed conflict and extractive industries undermine enforcement. Legal frameworks still support economic exploitation.
India – Ganges & Yamuna Rivers	Granted legal personhood, equated to a minor incapable of exercising rights independently – in need of a curator (High Court of Uttarakhand decision).	The state, through three appointed officials, acts as the river's guardian under the *parens patriae* legal doctrine.	Recognized through public interest litigation brought by a renowned lawyer, concerning environmental degradation. Rooted in Hindu cultural and religious significance, combined with post-1950 secular environmental laws.	No specific rights granted beyond legal personhood. However, it provides a stronger basis for enforcing environmental protections.	Jurisdictional complexity, cultural-religious conflicts and enforcement challenges due to governance across multiple states.

Common Themes:

- Legacy of colonization leading to marginalization of Indigenous/ethnic/local communities, limiting their influence on environmental decisions.
- Implementation complexities due to institutional and legal gaps, extractivism and economic reliance on natural resources.
- Challenges in integrating ancestral values with Western legal principles for effective environmental governance

and cultural rights of local populations. The ruling mandates a commission of guardians and an expert panel to oversee implementation, combining state oversight with local participation. However, challenges related to institutional capacity, enforcement mechanisms and ongoing extractive activities have raised questions about whether this model can effectively ensure the protection of the river and its surrounding communities. While these cases share common influences, they also reveal how legal transplants do not occur in a vacuum but are mediated by the legal, cultural and institutional frameworks of each jurisdiction. The idea of granting legal personhood to rivers has travelled across legal systems – most notably between Aotearoa New Zealand, India and Colombia – but has been transformed through interaction with pre-existing legal structures and socio-political realities. This underscores the transnational dimensions of legal pluralism, where global legal ideas are reinterpreted, contested and embedded within specific local contexts.

In India, the Uttarakhand High Court's 2017 decision granted legal personhood to the Ganga and Yamuna rivers, drawing on Hindu religious traditions and the legal status of deities as juridical persons. The ruling invoked the *parens patriae* doctrine, assigning the state the role of guardian to represent the rivers' interests. Unlike the Aotearoa New Zealand and Colombian cases, this decision did not create new mechanisms for community environmental participation and was appealed before being stayed by the Supreme Court, meaning it has not been implemented in any way. Additionally, jurisdictional challenges related to the transboundary nature of the Ganga River raise questions about how such a model can be applied across multiple governance structures.

These different approaches highlight ongoing debates about the role of legal personhood as a mechanism for environmental protection. In Hohfeldian terms, recognizing nature's rights goes beyond establishing that humans have duties towards ecosystems; it means that rivers such as the Atrato, Whanganui, Ganga and Yamuna have claims to protection, requiring state and community actors to safeguard them. In Colombia, for instance, the state and local communities share the obligation to ensure the Atrato River's protection and restoration, positioning nature's rights as a claim-right that corresponds to human duties of care. In contrast, Aotearoa New Zealand's Te Awa Tupua model emphasizes liberties, affirming the river's right to exist as a living whole and flow naturally, while also introducing powers that enable Māori collaborative governance. Unlike the Colombian case, where the Atrato River cannot own property or engage in transactions, the Whanganui River has the legal capacity to do so, raising further questions about the scope and practical implications of granting nature such rights.

India's case introduces an alternative model based on power and immunity, as the Ganga and Yamuna rivers are assigned state-appointed guardians who can act on their behalf. However, this approach does not necessarily create new participatory mechanisms for local communities; rather, it places decision-making authority primarily in the hands of the government, leaving the rivers under exclusive state oversight. Additionally, the Te Awa Tupua Act not only grants the Whanganui River rights and powers but also states that it will hold the "duties and liabilities of a legal person", an aspect that has been subject to debate. This raises broader legal concerns about whether natural entities should bear legal responsibilities like those of corporate or human persons, illustrating the complexity of these legal frameworks and their potential consequences for environmental governance.

The diverse models of legal personhood for nature raise broader questions about the role of law in environmental governance, the limits of state-centric legal frameworks and the recognition of non-state legal orders. While proponents argue that granting legal rights to rivers provides a necessary tool for addressing ecological degradation, others caution that these frameworks risk becoming symbolic without robust enforcement mechanisms. The case studies analysed illustrate how legal pluralism shapes these debates by foregrounding the interactions between state law and Indigenous and Afro-descendant legal traditions, as well as by demonstrating how transnational legal borrowings are reinterpreted through local normative orders.

While proponents argue that granting legal rights to rivers provides a necessary tool for addressing ecological degradation, others suggest that these legal personhood models risk becoming symbolic rather than enforceable. In all three cases, legal recognition alone has not resolved key governance challenges, particularly when rivers span multiple jurisdictions or where existing resource management frameworks remain intact despite constitutional innovations. This raises broader philosophical and legal questions about whether nature's rights should be conceptualized as individual claims, collective rights, state responsibilities or other, and how they should be enforced in practice. Some perspectives emphasize the potential of collective rights approaches, such as Colombia's biocultural framework, to create legal structures that recognize interdependencies between human and ecological systems. Others focus on the importance of enforcement, examining whether existing legal structures provide sufficient mechanisms to ensure that nature's legal rights translate into practical protections.

The recognition of rivers as legal subjects in Aotearoa New Zealand, India and Colombia illustrates multiple legal pathways for integrating the agency of nature into legal systems. These cases raise questions about how different legal traditions, governance structures and cultural perspectives shape the implementation and effectiveness of nature's legal personhood. As legal systems continue to experiment with rights of nature frameworks, the extent to which these models influence broader environmental governance structures, state obligations and legal standing for ecosystems remains an area for further inquiry.

Notes

1 This conception of the rights of nature, in which humanity is once again placed at the centre of the legal universe (this time as a protective parent or guardian) can aptly be described with the words of Anna Grear, when she speaks about the need

for legal transformation to confront the planetary crisis: "One thing seems certain: if the law is to respond to the multiple crises afflicting the Earth, and if rights are to be deployed, we need to get rid of the notion of a rights-bearer who is an active, wilful human subject, set against a passive, acted-upon, nonhuman object. The law, in short, needs to develop a new framework in which the human is entangled and thrown in the midst of a lively materiality – rather than assumed to be the masterful, knowing centre, or the pivot around which everything else turns" (Grear 2019).

2 This risk is well illustrated by the analogy Kauffman draws between blood in the human body and water in a river basin: "While it may sound odd to think of rivers (understood as watershed ecosystems) holding water rights and selling water as property, this is conceptually no different than people selling their blood to blood banks" (Kauffman and Martin 2021, 229).

3 Following Jones (2022) collective rights, or group rights, are held by a group as a whole rather than by individuals. Unlike individual rights, which focus on personal claims to freedom, collective rights pertain to goods that are shared and public to the group's members. Denise Réaume (1988) suggests that group rights often relate to "participatory goods"– goods that require collective participation, such as culture, language or social activities. These goods are inherently social and cannot be fully enjoyed without the involvement of others, meaning they must be framed as group rights. Ultimately, the distinction between individual and collective rights depends on both the nature of the goods and the ethical considerations surrounding their shared enjoyment (Morauta 2002, Réaume 1988, in Jones 2022).

4 The Inter-American Court of Human Rights has ruled in this sense in Advisory Opinion 23 of 2017.

5 Early foundational contributions by John Griffiths (1986) and Sally Engle Merry (1988) laid the groundwork for scholarly debates on legal pluralism, while more recent analyses by Ralf Michaels (2009) B. Tamanaha (2021, 2017, 2011) developed the discussion further, particularly in relation to globalization. Paul Schiff Berman's *Oxford handbook of global legal pluralism* (Berman 2020) consolidates contemporary perspectives, illustrating the diversity of approaches in the field.

References

Barbaras, Renaud. 2019. *L'appartenance. Vers Une Cosmologie Phenomenologique.* Peeters.

Berman, Paul Schiff, ed. 2020. *The Oxford handbook of global legal pluralism.* Oxford University Press. doi:10.1093/oxfordhb/9780197516744.001.0001.

Bonilla Maldonado, Daniel. 2006. *La constitución multicultural.* Siglo del Hombre Editores.

Cantillo Ramírez, JhonFredy, Juan CarlosLópez Arboleda and Jorge Iván Hurtado Bonilla. 2019. "El reconocimiento del río Atrato, sus cuencas y sus afluentes como sujetos de derechos." http://repository.unilibre.edu.co/handle/10901/19919.

Cartabia, M. 2010. "The age of new rights." [working paper]. https://www.boa.unimib.it/handle/10281/67794.

Castañeda Ruiz, Hugo Nelson, Ángela María Gómez Osorio, Helena Pérez Garcés and Jhonny Alexander Herrera Mejía. 2019. "La declaratoria del Río Atrato como entidad sujeto de derechos: una oportunidad para la construcción de un proyecto presente-futuro de territorio sustentable." *Revista Kavilando* 11 (2): 417–433.

Congreso de la República de Colombia. 1993. "Ley 70 de 1993: Por la cual se desarrolla el artículo transitorio 55 de la Constitución Política." https://www.funcionpublica.gov.co/eva/gestornormativo/norma.php?i=7388.

Constitutional Court of Colombia. 2016. Judgment T-622. Atrato River Case.

Cribb, Miriama, Elizabeth Macpherson and Axel Borchgrevink. 2024. "Beyond legal personhood for the Whanganui River: Collaboration and pluralism in implementing the Te Awa Tupua Act." *The International Journal of Human Rights* (February 16): 1–24. doi:10.1080/13642987.2024.2314532.

Cruz Hermida, Giselle de la, Ana Elizabeth Ramírez Gómez and Rafael Enrique Valenzuela Mendoza. 2021."El debate contramayoritario: Aproximaciones teóricas en Estados Unidos e Hispanoamérica." *Revista chilena de derecho y ciencia política* 12 (1): 57–77.

Cuesta Mosquera, Arlinton, and Catalina Vallejo Piedrahíta. 2024. "Take me to the River: Have riverine rights enhanced community participation in environmental governance within the Atrato River basin?" *The International Journal of Human Rights* (July): 1–19. doi:10.1080/13642987.2024.2367468.

Gewirth, Alan. 1986. "Why rights are indispensable." *Mind* 95 (379): 329–344.

Gómez Erazo, Alejandra. 2012. "La Ley 70 de 1993 y la participación política de las comunidades negras." *Trans-pasando Fronteras*2 (December): 31–45. doi:10.18046/retf.i2.1321.

González-Serrano, María Ximena. 2024. "Rights of nature, an ornamental legal framework: Water extractivism and backbone rivers with rights in Colombia." *The Journal of Peasant Studies* 52 (2): 322–342. doi:10.1080/03066150.2024.2349228.

Grear, Anna. 2019. "It's wrongheaded to protect nature with human-style rights." *Aeon* (blog), 19 March. https://aeon.co/ideas/its-wrongheaded-to-protect-nature-with-human-style-rights.

Griffiths, John. 1986. "What is legal pluralism?" *The Journal of Legal Pluralism and Unofficial Law* 18 (24): 1–55. doi:10.1080/07329113.1986.10756387.

Herstein, Ori. 2023. "Legal rights." In *The Stanford Encyclopedia of Philosophy*, edited by Edward N. Zalta and Uri Nodelman. Metaphysics Research Lab, Stanford University. https://plato.stanford.edu/archives/win2023/entries/legal-rights/.

Hohfeld, Wesley Newcomb. 1917. "Fundamental legal conceptions as applied in judicial reasoning." *The Yale Law Journal* 26 (8): 710–770. doi:10.2307/786270.

Jones, Peter. 2022. "Group rights." In *The Stanford encyclopedia of philosophy*, edited by Edward N. Zalta and Uri Nodelman. Metaphysics Research Lab, Stanford University. https://plato.stanford.edu/archives/fall2022/entries/rights-group/.

Kahui, Viktoria, Claire W. Armstrong and Margrethe Aanesen. 2024. "Comparative analysis of rights of nature (ron) case studies worldwide: Features of emergence and design." *Ecological Economics* 221 (July), 108193. doi:10.1016/j.ecolecon.2024.108193.

Kauffman, Craig M., and Pamela L. Martin. 2021. *The politics of rights of nature: Strategies for building a more sustainable future.* MIT Press.

Kurki, Visa A.J. 2019. "A short history of the right-holding person." In *A theory of legal personhood*, edited by Visa A.J. Kurki. Oxford University Press. doi:10.1093/oso/9780198844037.003.0002.

Kymlicka, Will. 2024. "Rethinking human rights for a more-than-human world." In *More than human rights: An ecology of law, thought and narrative for earthly flourishing*, edited by César Rodríguez-Garavito. New York University Law MOTH Project.

Macpherson, Elizabeth Jane. 2020. *Derechos constitucionales, derechos humanos, derechos indígenas: el lado humano de los derechos de la naturaleza*. Universidad Externado de Colombia. https://bdigital.uexternado.edu.co/entities/publication/4d21c943-6ac2-487d-a560-6b9e3231ff5c.

Macpherson, Elizabeth, Axel Borchgrevink, Rahul Ranjan and Catalina Vallejo Piedrahíta. 2021. "Where ordinary laws fall short: 'Riverine rights' and constitutionalism." *Griffith Law Review* 30 (3): 438–473.

Martin, Pamela L., and Craig M. Kauffman. 2019. "How courts are developing river rights jurisprudence: Comparing guardianship in New Zealand, Colombia, and India." *Vermont Journal of Environmental Law* 20: 260–289.

McNeish, John-Andrew, and Jenny Moreno Socha. 2024. "Muddy waters: On the problematic political ecology of the Atrato ruling, Colombia." *The International Journal of Human Rights* 0 (0): 1–21. doi:10.1080/13642987.2024.2324121.

Melo-Ascencio, Diego Alejandro. 2024. "A postcolonial Marxist critique of the River-as-subject: Situating the Atrato River in its development aporias." *Antipode* 56 (5): 1754–1774. doi:10.1111/anti.13052.

Merry, Sally Engle. 1988. "Legal pluralism." *Law & Society Review* 22 (5): 869–896.

Michaels, Ralf. 2009. "Global legal pluralism." *Annual Review of Law and Social Science* 5 (1): 243–262. doi:10.1146/annurev.lawsocsci.4.110707.172311.

Morauta, James. 2002. "Rights and participatory goods." *Oxford Journal of Legal Studies* 22 (1): 91–113. doi:10.1093/ojls/22.1.91.

Mosquera Perea, Yosserth Yairth and Jessika Suleidy Rentería Bonilla. 2024. "Caracterización y marco evolutivo de la Ley 70 de 1993 y demás normas conexas en el marco del reconocimiento de los derechos de comunidades negras, afrocolombianas, raizales y palenqueras en Colombia: Balances y perspectivas (1993–2021)." http://hdl.handle.net/11396/8129.

Natarajan, Usha, and Julia Dehm. 2019. "Where is the environment? Locating nature in international law." *TWAILR: Third World Approaches to International Law Review.* https://twailr.com/wp-content/uploads/2019/08/Natarajan-Dehm-Where-is-the-Environment_Locating-Nature-in-International-Law.pdf.

O'Donnell, Erin. 2020. "Rivers as living beings: Rights in law, but no rights to water?" *Griffith Law Review* 29 (4): 643–668.

Pérez Caballero, Alfredo José, Zilath Romero González, Elfa Luz Mejía Mercado and Sandra Lorena Hidalgo Arango. 2024. "Efectividad de la ley 70 en el reconocimiento de derechos y la construcción de ciudadanía afrocolombiana." *Academia & Derecho* 17 (29): 1–16.

Raz, Joseph. 2006. "On the nature of rights." In *Theories of rights*, edited by C.L. Ten, 39–59. Routledge. doi:10.4324/9781315236308-3.

Réaume, Denise. 1988. "Individuals, groups, and rights to public goods." *University of Toronto Law Journal* 38 (1).

Rodríguez-Garavito, César. 2024. "Introduction & more-than-human rights: Law, science, and storytelling beyond anthropocentrism." In *More than human rights*, 15–20. NYU MOTH Project.

Salazar Ugarte, Pedro. 2011. "La suplantación del legislador democrático." In *Reforma y control de la constitución: Implicaciones y límites*, edited by Córdova Lorenzo and Cesar Astudillo, Serie Memorias, no. 3, 195–202. Universidad Nacional Autónoma de México.

Sánchez Jaramillo, Johana Fernanda. 2022. "Colombia: La naturaleza como sujeto de derechos entre el activismo y la contención." *Novum Jus* 16 (3): 189–218. doi:10.14718/novumjus.2022.16.3.8.

Sánchez Zapata, Diana Carolina. 2022. "El reconocimiento de la naturaleza como sujeto de derechos: Una oportunidad para repensar la planeación del ordenamiento territorial como función administrativa." *Revista Derecho del Estado*54 (November): 87–131. doi:10.18601/01229893.n54.04.

Sieder, Rachel. 2019. "Legal pluralism and fragmented sovereignties: Legality and illegality in Latin America." In *Routledge handbook of law and society in Latin America*, 51–65. Routledge. https://www.taylorfrancis.com/chapters/edit/10.4324/9781315645193-4/legal-pluralism-fragmented-sovereignties-rachel-sieder.

Sousa Santos, Boaventura de. 1995. *Toward a new common sense: Law, science, and politics in the paradigmatic transition*. Vol. 111. Routledge. https://api.law.wisc.edu/repository-pdf/uwlaw-library-repository-omekav3/original/24ba59d5f7586b5bb541de2b0fd5bcc14bd2d6c9.pdf.

Stone, Christopher D. 1972. "Should trees have standing?: Toward legal rights for natural objects." *Southern California Law Review* 45: 450.

Tamanaha, Brian Z. 2011. "The rule of law and legal pluralism in development." *Hague Journal on the Rule of Law* 3 (1): 1–17.

Tamanaha, Brian Z. 2017. "Understanding legal pluralism: Past to present, local to global." In *Legal theory and the social sciences*, 447–483. Routledge. https://www.taylorfrancis.com/chapters/edit/10.4324/9781315091891-17/understanding-legal-pluralism-past-present-local-global%E2%80%A0-brian-tamanaha.

Tamanaha, Brian Z. 2021. "Legal pluralism across the Global South: Colonial origins and contemporary consequences." *The Journal of Legal Pluralism and Unofficial Law* 53 (2): 168–205. doi:10.1080/07329113.2021.1942606.

Thomson, Judith. 1990. *The realm of rights*. Harvard University Press.

Twining, William. 2006. "Diffusion of law: A global perspective symposium on legal transplants." *Journal of Comparative Law* 1 (2): 237–260.

Twining, William. 2012. "Legal pluralism 101." In *Legal pluralism and development*, edited by Brian Z. Tamanaha, Caroline Sage and Michael Woolcock, 1st ed., 112–128. Cambridge University Press. doi:10.1017/CBO9781139094597.011.

Twining, William. 2021. "Normative and legal pluralism: A global perspective." In *The Oxford handbook of transnational law*, edited by Peer Zumbansen, 31–66. Oxford University Press. https://books.google.com.co/books?hl=en&lr=&id=reMmEAAAQBAJ&oi=fnd&pg=PP1&dq=william+twining+oxford+handbook+on+transnational+law&ots=_eFXsPTYyT&sig=eP-gfkUPFLw_7tLXNfrR_2CQEO4&redir_esc=y#v=onepage&q=william%20twining%20oxford%20handbook%20on%20transnational%20law&f=false.

Upadhyay, Kavita, and Bibhu Prasad Nayak. 2024. "Tracing the legal journey of petitions in the Uttarakhand High Court that became springboards for rights of rivers and nature in India." *The International Journal of Human Rights* 0 (0): 1–20. doi:10.1080/13642987.2024.2380850.

Waldron, Jeremy. 1993. *Liberal rights: Collected papers 1981?1991*. Cambridge University Press.

Wellman, Carl. 1995. *Real rights*. Oxford University Press.

Wesche, Philipp. 2021. "Rights of nature in practice: A case study on the impacts of the Colombian Atrato river decision." *Journal of Environmental Law* 33 (3): 531–555. doi:10.1093/jel/eqab021.

Wilson, Edward O. 2021. *Every species is a masterpiece*. Penguin.

Zillich, J.A. 1972. "Toxicity of combined chlorine residuals to freshwater fish." *Water Pollution Control Federation* 44 (2): 212–220.

7 The politics of riverine rights

Axel Borchgrevink

Introduction

The riverine rights cases we examine in this book – the legal personhood for the Whanganui River in Aotearoa New Zealand, for the Atrato River in Colombia and for the Ganga (Ganges) and Yamuna rivers in India – have all come about through decisions in state institutions empowered to make and uphold formal law; the courts in Colombia and India, Parliament in Aotearoa New Zealand.[1] Thus, they are part and parcel of the formal legal system of their countries. The very notion of legal rights presupposes the existence of the state and its institutions, including the law. In one – idealized and ideal-typical – understanding of the law in modern states, it emerges from a government representing the people, as a binding set of rules that prescribes how society will function, with the objective of rational governance to the benefit of the nation. Of course, this is not the full story. Laws are embedded in society. They arise out of complex processes, they reflect as well as regulate power constellations, and their effects are mediated in the interplay of conflicting interests, actors and institutions.

As the previous chapter explains (Vallejo Piedrahíta, Montoya-Restrepo, Arango Duque and Gaviria Mira this volume), notions of river rights may sit uneasily alongside other laws and regulations. Moreover, these cases can be understood as instances of legal pluralism, where Indigenous and local understandings and principles are integrated into state law. To understand why and how these cases came about, and to appreciate the effects they have or may come to have, they must be situated within their social and political context. This chapter seeks to do this for our three cases and gives a comparative perspective on key dimensions. This juxtaposition allows the great differences between our cases to be explored in relation to the different cultural, social, institutional and political situations of their respective countries. But it also brings out similarities and lessons to be drawn.

These cases show how the legal system, and the legal decisions and mechanisms, are embedded in a wider political context, and the dynamics can only be understood in relation to that context. Yet, as the review of the cases demonstrates, the context and the dynamics are widely different between

DOI: 10.4324/9781003581208-7

them. This section seeks to understand the mechanisms and structures that explain these differences – while also pointing out the parallels that exist. The chapter offers three comparative lenses through which to analyse how context shapes riverine politics: the characteristics of the state, the political economy of extraction and resource use, and the politics of knowledge. Viewing the three cases through these lenses brings out contrasts and commonalities, and points towards more general mechanisms and conclusions.

Postcolonial states and the politics of sovereignty

One issue that immediately appears from the juxtaposing of these cases is the great difference between the three states. Obviously, there are differences in scale – from India with a population of more than 1.4 billion, to Colombia's 52 million and Aotearoa New Zealand's 5.2 million. But qualitatively, there are also huge differences: Aotearoa New Zealand is a welfare state, with a strong economic base, a well-functioning bureaucracy and political system, and only limited challenges to law and order. Colombia has a very long history of political violence, and there are large areas of its territory over which non-state armed actors continue to exercise de facto control, and for these reasons there are observers that have characterized the country as "a failed state". India is a huge, federal state, whose immense size and social and cultural complexity defy simple ways of characterizing it, save for pointing out this complexity as a defining characteristic. One common characteristic between the countries is that they are all former colonies. But still, colonial history, the process through which independence was achieved, and postcolonial experience are very different for these countries. While the colonial past is fundamental for understanding each of these cases, that past is so different that the consequences it has for our cases must be examined individually.

How then should we analyse the state for the purpose of comparing our three cases of rivers with personhood? Approaches to studying the state show tremendous variation and may be based on contradictory premises. The state can be understood as a political actor, as the prize in the political contest or as the arena within which politics take place. The state can be understood as a coherent, unitary and rational actor; as a set of interacting institutions that do not always operate smoothly together and that are susceptible to elite capture; or the concrete existence of the state may be challenged by alternatively seeking to grasp it as a set of effects, as a process or a relation or set of everyday practices, or a discursive construction legitimizing certain forms of knowledge (and delegitimizing others). The objective in this chapter is not to argue for a specific way of understanding the state. The objective here is simply to throw comparative light on the three cases. In the following, different approaches are drawn upon that bring out dimensions of our cases that facilitate comparison.

An obvious contrast between the cases is the difference in terms of the states' ability to carry out its policies and the effectiveness of the state apparatus, both for maintaining control over its territory and in terms of providing services. There is a large literature on state capacity and state strength, of relevance to understanding our cases. An important strand of this literature focuses on former colonial states. "Weak", "collapsed" and "failed states" have been defined by their lack of ability to provide services or to uphold law, order and a monopoly of violence (Rotberg 2003). In my view, this terminology is problematic because it measures actual states against a standard to which they are bound to fall short. In short, they are defined by what they lack (Eriksen 2010). For our purposes, labels such as weak, collapsed or failed have no use. But this literature's focus on the capacity of states to impose its policies throughout its territory – the degree to which it has de facto internal sovereignty – is an important dimension when comparing our cases. It is evident that part of the reason why legislation in Aotearoa New Zealand holds considerable promise of a positive impact, while what at first sight appears a similar legal decision in Colombia has had little impact, has to do with state control over its territory. And it is equally clear that for any socio-legal study, the degree to which a state can exercise its authority and make its laws count – that is, the degree to which the state has de facto (internal) sovereignty – must be a crucial dimension.

Hansen and Stepputat argue that when the focus is on de facto sovereignty – "the ability to kill, punish and discipline with impunity", then it need not be "grounded in formal ideologies of rule and legality" (2006, 296). In this perspective, sovereignty is not a prerogative only of the state; one will often find different types of sovereignty that coexist and overlap. In particular, they claim, this was a characteristic of the colonial state, where sovereign power was typically fragmented and distributed among different informal forms of local authority. For maintaining control over the territory, the formal apparatus of the colonial state often relied on the power of traditional authorities such as chiefs, as well as on trading companies, missionaries, planters and concessionaries from the colonial power, who were often given wide-ranging powers to uphold order with high levels of violence, in the tacit or explicit name of the colonial state. Moreover, they claim, while independence did mean changes to authority and sovereignty, the characteristic of multiple, fragmented, nested, layered and overlapping sovereignties has carried over and continued, even if changing global political and economic structurers are bringing new power wielders to the fore: organized crime, private security companies, extractive corporations, insurgents and vigilante groups (Hansen and Stepputat 2006). McNeish draws on these ideas for the concept of "resource sovereignty", which points to the politics of the struggle over control over natural resources between states, corporations, local communities, illegal actors and others in the context of extractivism (Logan and McNeish 2012, McNeish et al. 2015, McNeish 2021).

Eriksen (2010) makes a parallel point about sovereignty and postcolonial states. He does this by focusing on the relationship between state and society. Eriksen is concerned with the effects of the norm of the idealized modern state, developed with reference to European experiences. This norm, he suggests, is a pervasive idea with profound effects upon the world through its influence on practice and the way it is used to legitimize power. Yet it is an idealized idea, far removed from the reality of most existing states, and therefore of little use as an analytical category. This idea of the state holds that the sovereign state represents society as a whole, and that its policies are on behalf of society's common interests. State is separate from society yet represents it. Moreover, there is the implication that the state has control over its territory, and the ability to carry out its policies. Eriksen points out that the colonial state, while importing the institutions of the European state, was clearly not the sovereign expression of the will of its inhabitants. On the contrary, it was established to serve the interests of the colonial power. While independence changed this, it did not automatically establish the relationship between state and society of the idealized idea of the state. Eriksen – perhaps mainly with reference to the experience of African states – points to mechanisms that led to a kind of inversion of the relationship between state and society in postcolonial cases. Instead of representing society as a whole, the state will often be captured by specific interests. And instead of maintaining the Weberian separation between state and society, neopatrimonial logics blur this distinction.

Writings on settler colonialism have focused on how it is radically different from other forms of colonialism. As Veracini writes, while all colonialism is about foreign domination, it is not the same if that domination comes in the form of "You, work for me!" or in the form of "you, go away! [so that I can take over your land]" (Veracini 2011, 1). This is of course an analytical distinction, and in practice there will be elements of both forms in all colonialism. Still, in a country like Aotearoa New Zealand, it is clear that displacing the Māori and taking over their land has been the dominant form that colonization has taken. The distinction between colonialism and settler colonialism has important implications for how the postcolonial is understood. To quote Carey and Silverstein:

> Where the 'post' in post-colonialism refers to the ongoing effects of colonial rule in states that have been formally decolonized, settler colonial studies consider those political and geographic contexts in which the colonisers never left.
>
> (Carey and Silverstein 2020, 1)

Silverstein's phrase that there is "nothing post about colonialism in Australia" (Silverstein 2016, quoted in Carey and Silverstein 2020, 3) could equally well be applied to Aotearoa New Zealand.

These perspectives are relevant for our three cases, which all concern postcolonial countries. As stated, the colonial and postcolonial experiences differ highly among them.

The Whanganui River

The politics surrounding river rights in Whanganui are rooted in the struggles over sovereignty that began with the *Tiriti*/Treaty of Waitangi in 1840, marking the start of British colonization in Aotearoa New Zealand. The *Tiriti* promised Māori chiefs *te tino rangatiratanga* – authority, chieftainship – over their lands and treasured possessions. However, the growing presence of British settlers, backed by the Crown and The New Zealand Company, soon overshadowed Māori authority. By the end of the 19th century, little more than fifty years after the treaty was signed, the Māori were a small minority, dispossessed of land and disenfranchised (King 2003, 168–188, 240–258).

Māori sovereignty over the Whanganui River was never ceded but consistently claimed by the Whanganui *iwi* (tribe), though the government increasingly controlled the river. From 1860 to 1960, the government confiscated land for the Whanganui Port, removed obstacles to steamboat navigation (destroying valuable eel weirs), licensed gravel extraction, placed environmental protections on Māori-owned riverbanks, attempted to exterminate native eels while introducing foreign species and licensed the diversion of the river's headwaters for hydroelectric power. These decisions were made without heeding Māori objections (Waitangi Tribunal 1999, Bayley 2007).

Despite this, Māori continuously resisted through court claims, petitions and civil disobedience. Court cases were often lost, and petitions ignored. The most significant legal battle, the riverbed litigation (1938–1962), ultimately concluded that while Māori ownership of the riverbed after 1840 was not in dispute, it had been extinguished by a Native Land Court decision in the late 19th century (made without Māori involvement). Thus, from 1840 to the 1960s, Māori sovereignty was usurped by the Crown and settlers through a combination of numerical superiority, resources, force and control over the state (Waitangi Tribunal 1999).

For Māori, the independence of the settler colonial state from the United Kingdom implied no changes to this state of affairs. The process of achieving formal sovereignty as a state was a gradual affair, spanning 90 years (1857–1947). Full independence for the Aotearoa New Zealand state from the UK came without struggle, and according to King, was only reluctantly ratified by Aotearoa New Zealand's parliament after the second World War (King 2003, 422).

However, the emergence of an activist Māori rights movement in the 1970s brought Treaty of Waitangi grievances into national debate. While colonialism may still be ongoing in settler colony countries, it must nevertheless be understood as having different phases. The establishment of the Waitangi Tribunal in 1975, particularly after 1985, when it was empowered to investigate claims dating back to 1840, marked a shift away from unidirectional colonization and the dispossession of Māori. (Although critical scholars such as Mikaere (2011, 139ff) and Tauri and Webb (2011) have interpreted this as simply a way to contain or control Māori claims.) By 2015, the Tribunal had registered over 2,500 claims and issued final reports on 123, leading to government negotiations with the different *iwi*. The objective was to settle grievances through apologies and reparations. By the end of 2024, 90 settlements had been signed (Te Arawhiti 2024), including the Whanganui River Claim Settlement of 2014, which led to the 2017 Te Awa Tupua (Whanganui River Settlement) Act. Recent political changes have affected this process. The current coalition government has enacted policies that reduce Māori rights and limit the Waitangi Tribunal's scope, which has led to strong Māori protests and a halt to negotiations over new settlements.

The Whanganui *iwi* had filed their claim to the river in 1990, and the Waitangi Tribunal's 1999 report concluded that the Treaty of Waitangi guaranteed their rights to ownership, authority and control of the river, rights that had never been willingly relinquished. Negotiations occurred between 2002 and 2004, and again from 2009 until the settlement was signed in 2014. The idea of legal personhood for the Whanganui River was a way to overcome the government's unwillingness to cede ownership of the river to Māori. With the Whanganui River becoming a legal person, it could own itself (Finlayson and Christmas 2021, Geddis and Ruru 2019).

The Whanganui River Claim Settlement required commitment and creativity from both parties. The settlement and legislation are unique in their reliance on *kawa* (Māori values or law), and the detailed administrative and management institutions created. While the government incorporated *kawa* into the Act, the Māori had to compromise on ownership and accept the continuation of existing private ownership of land along the river and established rights to the use of water. This includes the diversion of water for hydropower, despite its negative impact on the river.

The Act offers some protection against changing government policies, in requiring all management decisions to adhere to its principles. However, it shifts the political arena to regional and local levels. Local mobilization was necessary to ensure the District Council understood the need for new decision-making processes. Applying the Act's principles opens new political arenas, such as involving local people in discussions with authorities on programme implementation (Cribb et al. 2024).

Local level politics – as all politics – involve taking different interests into account. As the non-Māori population of the district is considerably larger than the Māori, one might expect divisions to arise along such lines. So far there are not any indications of this. One reason is the way that new institutional mechanisms involve the participation of all groups, Māori as well as non-Māori (Cribb et al. 2024, Cribb this volume). For decisions concerning the river, one could therefore imagine that dimensions that do not simply map onto ethnic differences, such as urban/rural, ownership or not of land along the river, involvement in agriculture, industry, tourism or others, etc., would be more relevant dividing lines for conflicting interests. But the experience up to now has not involved major cases of strong disagreements. There has been agreement on the principle of seeing the interests of the river as the interests of all.

While the Te Awa Tupua Act sets new parameters for how things are to be done, implementation will always be deeply enmeshed with politics. An interesting case for the future will be the negotiations over a possible new licence for the power company Genesis to continue its diversion of water from the Whanganui when its current concession expires in 2036. As this will involve entrenched economic outside interests, the outcome of those negotiations may be a good indicator of the true potential of the Te Awa Tupua Act to empower the Whanganui *iwi* to exercise their authority in a way that reflects their close connection with the river.

The Atrato River

Colombia achieved its independence from the colonial power Spain already in 1819 – two decades before the colonization of Aotearoa New Zealand. Independence came after a liberation war led primarily by representatives of Colombian-born descendants of the Spanish colonizers. Colombia thus became an independent state in an era that was very different from the post-World War II period when Aotearoa New Zealand – and India – became independent. While European-style institutions had been established in Colombia by the Spanish, it was a rudimentary state with limited reach throughout its territory. Processes of state formation have been dominated by elites descended from the former colonizers and their internal struggles for control over the state. From the viewpoint of Indigenous and Afro-descendant communities – such as the ones that today live along the Río Atrato – the replacement of the *peninsulares* (Spanish people born in Spain) by the *criollos* (descendants of Spaniards born in the colony) as the ruling elite had little significant effect (Randall 2015).

Throughout its two centuries of postcolonial existence, it is striking how far the Colombian state has been and still is from having a monopoly of violence throughout its territory. Partly as an effect of a state formation process dominated by the contest between elites seeking rents and control over the state and partly due to the difficult geography of the country, the state has had limited presence and reach, particularly in remote areas. As a result, Colombia has experienced extremely high levels of violence over its two centuries of existence (Coatsworth 2003). The Colombian state has never achieved internal sovereignty – control of law and order – throughout its territory. In the 19th century, Colombia experienced a long series of civil wars between conservatives and liberals, to some extent reflecting the different interests of large landowners and of an urban-based bourgeoisie. New periods of civil war between the main political parties followed from the mid-20th century, eventually morphing into complex armed conflicts involving the national army, revolutionary rebel groups, paramilitary organizations and organized crime (Gray 2008). While levels of violence have been reduced, especially over the past decade, armed groups and illegal economic activity still operate with impunity in parts of the national territory. Sovereignty is distributed between competing actors.

The Colombian state, though, is a perplexing mix of lack of sovereignty as described, with extremely high levels of violence, and yet a fairly stable system of governance. The country has achieved stable economic growth over the past 50 years, there is a large, educated middle class, elections are held regularly, with votes fairly counted and results widely accepted, the judicial system holds a strong position, and a representative civil society exists (Gray 2008, 74). The 1991 Constitution was seen as consolidating the liberal democracy of Colombia, and it has been called the ecological constitution for its many articles dealing with environmental protection (Cuesta Mosquera and Vallejo Piedrahíta 2024, 7). This Constitution established Colombia's Constitutional Court, which has played an important role in developing jurisprudence independently of the executive branch of government (Benítez-Rojas 2022).

The petition brought before Colombia's Constitutional Court by the Indigenous and afro-descendant communities living along the Río Atrato was primarily a reaction to the state's failure to assume and exercise sovereignty over this part of its territory. Río Atrato runs through the Chocó, a department rich in cultural and biological diversity. It is also a remote region, with little infrastructure, limited state presence, widespread poverty, a violent history and the presence of several types of illegal armed actors. With very little road infrastructure apart from the highway connecting the regional capital Quibdó with Medellín, the river continues to be crucial for transport for the *Chocoanos*. Most of them live in smaller villages along the river. The majority of these are black descendants of former slaves, but there are also communities of different Indigenous peoples.

Throughout its history, the extraction of gold from the riverbed has been a key activity for the economy of the region. Pre-Colombian Indian communities along the banks were engaged in artisanal gold mining, an activity the Spanish colonizers scaled up through the import of African slaves. In recent decades, gold mining has expanded with the use of modern machinery and toxic chemicals. There is a complex mix of legal, large-scale mining, illegal large- and medium-scale mining and small-scale traditional mining. The illegal mining is dominated by armed actors, often operating in collaboration with drug cartels and/or rebel guerilla groups. The extraction methods lead to large-scale destruction of ecosystems and the pollution of the river (McNeish and Moreno Socha 2024).

This contamination and the health problems for people living by the river were the main reasons for the petition filed before the Constitutional Court by the afro-descendant communities along the river. In its landmark ruling, the court declared the Atrato River to be a subject of rights and introduced the concept of biocultural rights. It also instructed the government to undertake a series of actions to halt the destructive mining and to clean the river. While the decision has given the communities of the river an improved position for dealing with the government, the concrete effects in terms of cleaning the river have not materialized. The reasons for this are many and include issues described above: the state's lack of sovereignty in regions such as Chocó, a fragmented state where coordination between ministries, agencies and other institutions is complex, a lack of funding and maybe also of political will to address these issues (Vallejo Piedrahíta, McNeish, Moreno Socha, Cuesta Mosquera and Alfonso León this volume; also McNeish and Moreno Socha 2024).

The Ganga and Yamuna rivers

In India, the colonial state was imposed upon already existing, complex political structures. While British colonialism had far-reaching effects on governance in the sub-continent through its introduction of new technologies of statecraft – British law and legal system, representation, measurement, sociology – subaltern historians have pointed out the limited relationship established between the colonial state and Indian society (Hansen 2019). "Dominance without hegemony" in Guha's phrase (Guha 1989). India's independence in 1947 came after decades of campaigning by a self-rule movement that was heterogenous but which had rootedness in Indian society as a common denominator. Thus, it not only came about at a very different time in global history from the independence of Colombia but was also the achievement of a very different type of movement.

The new state that emerged over the first decades after independence faced enormous challenges in integrating its immense cultural, social, religious, political and economic diversity and resisting the concomitant centrifugal forces. The Constitution of India (1950) established a secular, federal state based on universalistic principles. Federalism implied a divided, shared and negotiated sovereignty, different from the Western-unified sovereign state model (Rudolph and Hoeber Rudolph 2010).

In summing up the achievements of the first decades under Nehru, Khilnani posits that its great success lay in establishing the state at the core of Indian society – transforming it "from an alien object to one that aspires to infiltrate everyday lives of Indians" (1998, 41). But while the state had become ubiquitous, it was also characterized by its inefficacy – cumbersome red tape and an under-resourced and incompetent bureaucracy – and was prone to capture by local elites (Hansen 2019, 8). Widespread grumbling about corruption is common to this today. It has been argued that this is evidence of both dissatisfaction with the actual functioning of the state as well as of internalization of the universalistic and impersonal principles on which it ideally should operate (Gupta 1995, Parry 2000).

The Indian judiciary has played a key role in developing and consolidating governance and the rule of law. After the emergency period under Indira Gandhi (1975–77), the courts developed an activist role to hold government accountable and to establish new legal principles in different areas, including corruption, elections and the environment (Maté 2015). Important in this respect has been the public interest litigation mechanism introduced from the late 1970s, which relaxed requirements of legal standing for filing petitions and allowed judges considerable procedural leeway and independent initiative – excessively so, according to some (Chandra 2018). However, there are signs that under the current rule of the Hindu nationalist Bharatiya Janata Party (BJP), there has been a transformation of the Indian state towards a more corporative and less secular state, where there is also less independence for the judiciary (Sundar 2023, Nielsen and Nilsen 2024).

Giving legal personhood to the Ganga and Yamuna rivers was in many ways a result of the situation described above: an inefficient state and an activist judiciary. Also, the two rivers are symbolically important to and perceived as goddesses by many Indians. They both have their origins in Uttarakhand in the mountainous north of the country and continue into other states. The Ganga basin is huge, home to half a billion people. Both rivers are highly contaminated. Reacting to government inaction in response to previous judgments that required cleaning-up measures for the river, Uttarakhand High Court made the landmark ruling declaring the two rivers to be legal persons, with all the rights of a living person. This decision was stayed after it was challenged in India's Supreme Court and therefore remains suspended until the Supreme Court makes a final ruling on the case.

Sovereignty politics compared

The issue of sovereignty is central for each of the three cases, but in very different ways. In Aotearoa New Zealand, from the perspective of the Māori, the settler state has too much sovereignty – a sovereignty that has been usurped from the original inhabitants. The Whanganui claim can thus be seen as a challenge to state sovereignty, and the Te Awa Tupua Act as an effort to create a more inclusive state through the recognition of a fairly radical form of legal pluralism – a situation of nested sovereignty, where authority over the river is devolved to *iwi* and the people living along the river.

In Colombia, the central issue is that the state lacks de facto sovereignty over its territory. In particular, it lacks monopoly of violence over remote regions such as the Chocó and Río Atrato. In these areas, different armed and illegal actors wield considerable sovereignty, often in contest with each other or with the state. The case brought before the Constitutional Court was an effort to make the state take control over this area and exercise sovereignty, and the state's limited capacity to do so remains the main reason for why the court's decision has had limited effects so far.

In India, it could be said that limited sovereignty stems from an ineffective state with an ever-present but ultimately weak bureaucracy prone to capture by elite interests. Degradation of the Ganga and Yamuna rivers may be understood as one effect of this. The two legal cases brought before Uttarakhand's High Court, as well as the court decisions in these cases, were attempts to get the state to exercise real sovereignty. The present situation of the Ganga and Yamuna ruling – left in limbo awaiting a Supreme Court process in an indefinite future – can be seen as yet another example of a state lacking effective sovereignty. India being a federal state means that sovereignty is not unitary; it is in some respects distributed between the states. When Uttarakhand High Court – a state with less than 1% of the Indian population, and with only the upper reaches of the Ganga and Yamuna rivers within its territory – declares these rivers, along with their tributaries and connected waters, to be legal persons, the issue of jurisdiction and distributed sovereignty arises. It can easily be perceived as the tail trying to wag the dog.

The political economy of river rights

Another lens through which the politics of riverine rights can be grasped – though by no means unrelated to the character of the state and its institutions – is by examining the economic interests behind the use and extraction of natural resources related to these rivers. Shapiro and McNeish claim that we are now in an era of *hyper-extractivism*, defined by a mentality as well as an economy based on the extraction of natural resources. This "predation of natural resources" implies violence to nature and to people and is a global phenomenon that is connected through production chains and international financial flows. It is founded in a specific conception of nature as a resource to be exploited, a universalizing ontology underlying European modernity as it has developed over the last 500 years (McNeish and Shapiro 2021). With this emphasis on the underlying conception of nature as a resource to be used, the notion of "extractivism" expands. Beyond referring to activities like mining and logging, it also covers activities such as industrial agriculture and power generation. Some forms of tourism may also be included under the term.

The Atrato River

In the case of Río Atrato, the economy of the region is dominated by mining, by far the most economically important activity. Gold mining took place even before the advent of the Spanish but was greatly expanded in colonial times, with the use of slaves of both Indigenous and African descent. For a time, the Atrato region was the world's largest producer of gold (Richardson and McNeish 2021, 156). The current ethnic composition of communities along the river has been shaped by this history, as indicated by the afro-descendant population accounting for more than 80% of the people living in the Department of Chocó. Nationally, this figure is around 9% (DANE n.d.). Mining for gold and other minerals continues to be a main activity, alongside agriculture and forestry. Agriculture is mainly small-scale and traditional, although there are also plantations of sugar cane, bananas, coffee and other crops. Today's gold extraction is a continuation of the traditional economic activity of the region. But it has been greatly expanded in scale as well as with new technology. Gold mining is carried out in different scales, and both legally and illegally (Richardson 2020).

Mining has had severe effects on the river and the people living alongside it. Large tracts of rainforest are cut down and the earth excavated, leaving huge scars in the terrain and changing the river course through sedimentation. Particularly damaging to people and environment is the illegal mechanized mining, both through its liberal use of mercury and other toxic chemicals and through its large-scale destruction of vegetation, soil and riverbeds (Richardson and McNeish 2021). According to Tubb, one modern excavator can move in one week what it would take an artisanal miner a decade to shift (Tubb 2015). The previously clear water of the river has become muddy and contaminated. Mercury contamination has severe health effects for people living along the river, including miscarriages and impaired child development. Mercury levels 50 times higher than accepted by the World Health Organization has been found in people of the area. Reduced catches of fish and damaged drinking water are further effects (Kauffman and Martin 2021, Richardson and McNeish 2021).

The illegal mining often involves armed actors, with links both to organized crime and insurgent and paramilitary groups. The Chocó has long been an arena of Colombia's civil war, with some of the most brutal massacres taking place along the Atrato River. The river also serves as a transport route for the drug

Figure 7.1. A *dragón*, a dredge used by illegal gold miners on the Atrato River. Photo by Daniel Alfonso León.

smuggling, and gold is reported to be used by the cartels for money-laundering. The combination of the extraction of high-value products, lawlessness and the presence of armed actors leads to high levels of violence. Colombia is the country in the world where most environmental defenders are killed, and activists and community representatives in Chocó are acutely aware of the threats they face (Richardson 2020). It is widely believed that segments of local government and law enforcement are also implicated in these activities by turning a blind eye. The importance of gold mining for the local economy is easy to observe in the departmental capital and only large town of Quibdó, where stores selling equipment for mining are frequently encountered.

The Ganga and Yamuna rivers

The rivers Yamuna and, especially, Ganga, are of immense economic significance for India and for the half billion or so of Indians living in the river basin. Its water is used for transportation, energy generation, irrigation, fishing and human and animal consumption. Its religious significance involves large-scale pilgrimages to sacred sites along the rivers and is the mainstay of a flourishing tourist industry. As is the case of India itself, the scale is astonishing: reportedly, there are 940 hydropower dams and diversional barrages in the Ganga Basin (Economic Times 2018). Irrigation accounts for 90% of water use and serves more than 9 million hectares in Uttar Pradesh state alone (World Bank 2014). A network of waterways that include channels and sections of rivers offers economical transport, and its expansion is on the government agenda (Economic Times 2018). Industries of all kinds are conveniently located along the river. The Kumbh Mela festival – which takes place in several locations, the most important of which are on the Ganga and Yamuna rivers – reportedly attracts 400 million pilgrims, creating income for the area while leaving considerable imprints on the rivers and their surroundings.

These strong economic interests linked to the Ganga and Yamuna rivers have greatly contributed to their deteriorated environmental condition. Untreated sewage from millions of people that is directly discharged into the river leads to high levels of bacteria. Industrial effluents and waste from tanneries and other industries along the river further contribute to pollution of the river. This is compounded by runoff from agriculture and the widespread practice of open defecation. Dams, river diversions and canals reduce flows and cause sedimentation. This in turn leads to changes of the river course, concerns over water levels and reduced dilution of pollutants. The extraction of sand and gravel and other forms of mining add to these problems.

The two cases brought before the High Court in Uttarakhand both concerned economic activities deemed to be harmful for the conditions of the Ganga and respectively called for a halt to illegal mining of the riverbed and stone crushing for gravel extraction, and to the release of human waste into the river along with untreated effluents from a series of industries. The Uttarakhand High Court, in its ruling, took the wider picture into account when justifying its decision, by claiming that the rivers were losing "their very existence" (Mohd Salim v. State of Uttarakhand & Others 2017).

The Whanganui River

In the case of Whanganui, as was described earlier, the economic interests of settlers and the Crown, both during the colonial period and after independence, were the main cause of the dispossession of Māori from their ancestral river and traditional territory and for the considerable changes in land use that followed. The crucial role of the river as a transport route in the first phase of colonization, the continuing clearing of forests for agriculture, livestock and timber, the extraction of sand and gravel from the river and the diversion of water for hydropower production were the main ways in which the natural environment facilitated economic activities. The effects were three-fold. The process generated income for settlers and entrenched their position in the region, it dispossessed and marginalized Māori, and it led to environmental changes in the form of habitat and biodiversity loss and deterioration of water levels and quality. The extensive clearing of forests for pastureland for sheep and other livestock, including on steep hillsides, caused erosion and runoff that have muddied and contaminated the river and reduced fish and eel populations, while the diversion of water for hydropower purposes has reduced the flow of water and contributed to the deteriorated state of the river.

While the Te Awa Tupua (Whanganui River Settlement) Act is transformational in several aspects, it still contains a clause stating that existing property rights and economic activities are not affected by the legislation. It is perhaps understandable that the law could not go further, given the importance the use of land and water has for the economy of the region and the people living there. But even though the Te Awa Tupua Act proclaims the river to be indivisible from the mountains to the sea, this clause sets limits to the areas of river and land for which sovereignty can be returned to the Whanganui *iwi*. Moreover, it allows the continuation of the livestock production and the diversion of water to the neighbouring watershed for power generation, with the environmental effects referred to. So, also in the case of Whanganui, in several ways the most successful of the examples we have studied, entrenched economic interests limit the reach of the new legislation.

Political economy compared

In all the three cases, we see that the legal decisions of granting personhood to the rivers respond to issues resulting from economic activities related to use of natural resources. Likewise, extractive interests also affect the reach and impacts of legal personhood, even if it is difficult to establish exactly to what extent. The problems created by extractivism, and the ways in which extractivism shapes outcomes of the granting of personhood, vary across the cases, though.

In the case of the Atrato River, the colonial mining for gold pushed Indigenous peoples to remote areas upstream and brought slaves from Africa whose descendants established communities along other parts of the rivers. These marginalized groups have been heavily affected by illegal mining that is carried out in very harmful ways for people, the river and the surrounding environment. The linkages of this gold extraction to armed insurgent and paramilitary groups as well as organized crime has contributed to an extremely precarious situation. This state of affairs is closely linked to the limitations of state sovereignty described in the previous section. The Constitutional Court's decision to declare the river a legal subject with biocultural rights was meant to address both the critical human rights situation and the serious environmental damage. It is reasonable to assume that the economic value of gold extraction, its importance to armed actors beyond the reach of the state and the way the interests of illegal extractivism have managed to infiltrate state institutions are contributing factors to the fact that neither of these objectives have been reached so far.

In Uttarakhand in India, the High Court grounded its ruling on the degraded state of the Ganga and Yamuna due to contamination and mining, and the failure of the state to uphold its obligations to protect the rivers. Concerns for human rights or welfare were in this case not among the motives for the decision to award personhood. While the petition by the Uttarakhand state government to have the ruling overturned referred to the complexity of jurisdictional and implementation issues, it is a likely assumption that the state's opposition to the ruling also had to do with its dependence on income from industrial and extractive activities on the rivers, just as de Micheaux argues was the case in the state's opposition to and undermining of the previous case of the protective zone, which was declared by the National Ganga River Basin Authority in 2010 for the upper reaches of the Ganga (de Micheaux 2019, 193–195).

In the case of Whanganui, the primary objective of the Te Awa Tupua Act was to resolve historical injustices and address the Māori's claim to sovereignty over their river, a sovereignty that had been lost through dispossession by settlers and their economic interests. Thus, as we have argued elsewhere (Cribb et al. 2024), this is first and foremost a case of Indigenous rights, even if there are also environmental issues that are addressed through the giving of legal personhood and rights to the river. As the Act states that ongoing property rights and economic activities will not be affected, these economic interests were included in the settlement, thereby significantly limiting the reach of the legislation.

The politics of knowledge

The politics of riverine rights also has an ideational side: a struggle over how we should understand the world and our place in it, and what principles should guide our actions, as individuals and as societies. To quote Petel:

> [Rights of nature] serves as a mechanism for determining who, among humans, holds the authority to define 'nature', to prescribe its acceptable uses, to shape the development models, and to strike the right balance between social and environmental concerns. These deeply political questions should be at the heart of the RoN [rights of nature] discussion.
>
> (Petel 2024, 34)

An analysis of this politics can be framed in many ways. One could talk of the contest between different ideologies, or different philosophies, or different ethics (anthropocentric versus ecocentric), or different economic rationalities. One could put it as the logic of a colonially imposed system of statecraft (or economy) meeting a decolonial alternative, or as a contest between modernity and capitalism against other ways of conceiving the relationship between humans and their environment. In recent years, the concepts of ontologies and ontological difference have been frequently used to characterize the conflicting understandings of nature that are involved in the politics of rights for rivers, or for nature.

As this text wishes to engage with the academic literature of the field, it would appear natural to follow that example. However, "ontology" is a tricky concept. Firstly, it is used in different and inconsistent ways by different authors, often without specifying what precisely they mean by the term. This is one reason to be cautious of using it. Furthermore, I see problems with its use within social science, in particular with the more radical versions of the ontological turn. (This is not the place to enter into the debate over the ontological turn. The interested reader is referred to Graeber (2015) and Bråten (2016) for critiques I largely concur with.) I will therefore not use the concept of ontology in this text, except when referring to what others have written.

In 1995, Barth argued that since the concept of "culture" lends itself to reification, essentializing, exoticizing and homogenization, we should look for other concepts. He proposed that it should be replaced by "knowledge", which he held to offer several advantages: it points to people's direct engagement with the world; it acknowledges continuous variation; it alerts us to knowledge distribution, transactions and flux; and it opens us to various forms of knowledge and ways of knowing. Thereby, he claimed, the concept of knowledge brings to the fore dynamic questions related to variation, positioning, practice, exchange, reproduction, change and creativity, and can also work against the hegemony of one type of knowledge – such as "Western", scientific or anthropological knowledge – over other types (Barth 1995, 66;

see also Borchgrevink 2014). Believing that "knowledge" can offer similar advantages compared to "ontology", I will be using this term in the following. It should be emphasized that this notion of knowledge is wide and encompasses cosmology, morality and attitudes. Furthermore, "knowledge" is used without assessing the truth claims implied. In his book *Balinese worlds* (1993), Barth developed a sociology of knowledge around the notion of "traditions of knowledge". Traditions of knowledge have their particular social organization that structures the distribution of knowledge. There are particular types of experts; specific forms of transference, reproduction and change of knowledge; criteria of validation and relevance, etc. Social scientists can map such traditions and analyse their interplay in complex societies (Borchgrevink 2014). While space will not allow full analysis of this here, I believe it is useful to think in terms of such traditions of knowledge when analysing the politics of knowledge.

The Ganga and Yamuna rivers

Different sets of knowledge about nature and the Ganga River are interwoven with politics, Hindu nationalism, development and conservation efforts. Indian discourse and activism on environmentalism exhibit two sets of partially overlapping, partially opposed dimensions.[2] On the one hand, there is the dimension of ritual purity and impurity and its relation to physical cleanliness and contamination (Alley 1998, see also Brandshaug et al. this volume). And on the other hand, there is the contrast between environmental movements that are left-wing and promote the interests of the poor and disadvantaged, versus those movements that are right-wing and concerned with an untouched nature where the poor, the marginalized and the non-Hindus appear as threats to be removed (Blair 2023, Sharma 2023).

There have been several large-scale popular mobilizations and protests against the damming, mining and contamination of the Ganga. The campaign against the Tehri Dam – mainly on the Ganga tributary (or source) Bhagirathi – which lasted from 1972 to 2006 when the dam was completed, employed a variety of tools, including rallies, work blockages, hunger strikes, mass arrests, national and international media coverage and lawsuits (Blair 2023). The campaign initially drew on the Chipko movement that was active in the area. This was a social movement to protect the rights of local communities to their trees and forests, against the commercial logging promoted by the state. Sunderlal Bahuguna was a key figure of the Chipko movement, and he also became an influential spokesman for the Tehri anti-dam campaign (Sharma 2009). The broad alliance in the struggle against the Tehri Dam included local people threatened by displacement, secular environmentalists as well as Hindu-religious activists, and a broad range of sympathisers – the secretary of the Communist Party of India, The Indian National Trust for Art and Heritage and The Worldwide Fund for Nature were among the signatories of a petition to stop the construction. The campaign relied on arguments from natural science on the ecological risks posed by the dam, as well as on the social consequences of displacement and on the spiritual and aesthetical value of the river. However, the campaign shifted emphasis towards the end, as the Hindu nationalist organization Vishva Hindu Parishad became increasingly dominant within the campaign. Opposition to the dam then came to be couched more in terms of the centrality of the Ganga for the (Hindu) nation, and the ills of the dam were constructed as related to the perceived threats posed by outsiders – Muslims, communist China and the capitalist and secular West (Sharma 2009, Mawdsley 2005).

Partially overlapping the Tehri anti-dam campaign in time was the campaign to stop further dams in the upper part of Bhagirathi through the establishment of the protected area Gaumukh-Uttarkashi Eco-Sensitive Zone. From 2004, local opposition to new dams managed to reach the national agenda, eventually leading to the protective zone being declared in 2012. Also this time the arguments invoked against the construction of dams combined perspectives from ecology and natural sciences with Hindu ideas of the river as a sacred mother and goddess. Of crucial importance for the success of this campaign was the visibility it gained from the five hunger strikes of GD Agarwal, a former professor of environmental engineering at the prestigious Indian Institute of Technology turned ascetic and activist for the protection of Ganga (de Micheaux 2019).

The symbolic value of Ganga has also been important in the rhetoric of Prime Minister Narendra Modi and the BJP. When Modi launched his electoral campaign on the banks of the river in Varanasi in 2014, he drew on imagery from the same knowledge tradition. He said that Mother Ganga had called him, and that he felt like a son going back to his mother's lap, and he promised to clean up the river (The Guardian 2014). When assuming office after his party had won the national election, one of his early acts was to establish the National Mission for Clean Ganga, *Namami Gange*. While the programme has received considerable funding and has resulted in the closing down of a number of point-sources of pollution and the establishment of new sewage treatment plants, critics point to faulty implementation and loopholes in the scheme that limit overall impact and mean that improvement of the river is negligible (A. Ranjan 2024).

Sharma (2023) analyses the environmental policies and rhetoric of the Modi government and argues that they combine two main themes. On the one hand there is the idea of the unity of Hinduism, the nation and the territory, where Hinduism connotes a particular ecological wisdom, rivers and especially the Ganga have a prominent place as sacred nature, and those outside the Hindu nation – Muslims, Adivasi, Christians – represent threats of pollution (Ponnaganti 2024). Sharma characterizes this as right-wing ecology (Olsen 1999). On the other hand, there is a vison of industrial modernization based on renewable energy and green technologies for a sustainable and prosperous future. As Sharma points out, this combination contains contradictions and paradoxes, such as between Modi's claim of protecting rivers and his support of the highly controversial Sardar Sarovar Dam on the Narmada River (Sharma 2023, 106). There may be signs that the balance between these two themes have shifted since 2014, and that the developmentalist attitude has become dominant. When GD Agarwal in 2018 started a new fast for the Ganga, demanding the end to hydropower projects and sand mining and increased efforts to clean the river, the Prime Minister ignored his pleas. Agarwal died after 111 days of fasting (Kumar 2018, see also R. Ranjan 2024).

The issue of rivers in India, and of the Ganga in particular, evoke strong emotions, for many reasons. One of them is Hindu views of Ganga as sacred, a goddess and mother and a national river. Also, the river is widely perceived to be polluted by sewage, industry and mining, and its flow is disturbed and reduced by dams, canals and irrigation schemes. On the other hand, the Ganga has great importance for livelihoods, and for its potential for transport, power generation, industrial development and state building. Different traditions of knowledge imply different understandings of what the situation is and what is at stake. Alley shows how the perception of Hindu devotees of the ritual impurity of the Ganga River is different from – and in some respects diametrically opposed to – the views of natural scientists on river pollution, a difference she ascribes to differences of worldviews (Alley 1998). De Micheaux argues that different ontologies underlie the disagreements between hydropower proponents and Hindu-religious protesters in the conflict over the establishment of a protective area in the upper reaches of Ganga's tributary Bhagirathi. In the proponents' ontology, the river is understood "as an external and material entity, … a resource to be exploited, 'unless it is wasted into the sea'" (the last phrase a quote from an activist interviewed by de Micheaux). Within the Hindu-religious ontology, on the other hand, the river is both a material river and a goddess, and simultaneously neither an object nor entirely separate from the human (de Micheaux 2019, 200).

The political disagreements over policies for the Ganga described above clearly demonstrate that there are contrasting views of rivers, nature and the Ganga. Yet, we should be wary of reducing them to simply being the outcome of the meeting of two opposing worldviews (Alley 1998) or ontologies (de Micheaux 2019). Even from the brief summary I have given of these processes, it is clear that the positions of the involved actors are more complex than that. The Tehri anti-dam campaign brought together a broad range of actors, including people motivated by understandings based on religious ideas as well as those building their opposition to the dam on natural science perspectives, thus spanning the "worldviews" and "ontologies" described by Alley and de Micheaux. Among those that could be grouped as motivated by Hindu ideas, there are considerable differences, from attitudes emphasizing harmony with nature to those that focus on the purity of the Hindu nation and the dangers of outsiders, and from Gandhian non-violence to violent measures against non-Hindus. The activist GD Agarwal was both a recognized natural scientist and a dedicated believer willing to fast himself to death for the sake of the sacred Ganga. Modi's BJP government combines environmentalist ideas with modernist visions of large-scale industrial development. Simple dichotomies – such as "Hindu religious vs developmentalist ontologies" – are not particularly helpful for understanding the complex politics of knowledge involved here.

What is worth noting, though, is how disconnected the Ganga and Yamuna ruling of the Uttarakhand Hight Court appears to be from these political processes. While the judges draw on many of the same ideas and traditions of knowledge mentioned in this section (Borchgrevink and Upadhyay this volume), the idea of rights of rivers is little known and fails to mobilize the strong environmental civil society of Uttarakhand – and of India as a whole. Likewise, rights of the river – which in the ruling is underpinned with reference to Hindu values, and which one could imagine would resonate with the Hindu nationalism of the BJP, in power both in Uttarakhand and at Central level – has failed to connect with bureaucrats as well as politicians. Indeed, it was the BJP government in Uttarakhand that challenged the ruling in the Supreme Court. This disconnect from civil society, state and government – and from the political process itself – can be seen as an inherent risk of the type of judicial activism the Uttarakhand High Court ruling exemplifies. It may well explain why the ruling has been so quickly sidelined.

The Atrato River

In Colombia, there is also a complex situation of diverse knowledge traditions around the Atrato River. The Indigenous and afro-descendant communities along the river have their own ideas and values, intimately connected to their way of life and interactions with their natural surroundings. This contrasts with what they encounter from authorities and outside interests. In the words of a representative of the Embera Dodoma Indigenous people:

> We started to contest the idea of natural resources because others see it as a means to acquire, to commercialize. We do not see it this way, but as ancestral heritage to maintain culture … The Atrato River is our heritage and because of this we fight to maintain the river.
>
> (Quoted in McNeish and Moreno Socha 2024, 14)

The representatives went on to express that despite the cultural differences between them and the afro-descendants of the Atrato, they were allies in the fight due to the common interests in and dependence on the river. A representative of another Indigenous group, the Embera Katio, saw the environmental destruction brought on by extractivist activities as threatening the very survival of their own knowledge and culture:

> Our main concern has been the extraction of natural resources, particularly in mining. Another worry has been the pollution of rivers and streams. This has generated, to some extent, a cultural breakdown because our existence revolves around nature and the environment. As the forests are gradually destroyed, it isolates us from our natural surroundings, leading to an imbalance in the Indigenous population. This affects learning, knowledge, health, and education
>
> (Quoted in Cuesta Mosquera and Vallejo Piedrahíta 2024, 6)

McNeish and Moreno Socha describe a situation of "ontological complexity" that not only involves differences between afro-descendants and Indigenous communities but also between these communities and the authorities. There are also differences in understandings among representatives of public institutions, for instance in terms of whether to give priority to social or environmental issues (McNeish and Moreno Socha 2024).

The ruling of the Constitutional Court clearly meant to address this complexity by introducing the concept of biocultural rights, with the intention of giving priority to that cultural knowledge that is part of a lifeworld enmeshed with the river. This is well intentioned and gives the communities a legal basis for making claims. Yet, in practice, the implications of this innovation remain limited, as the actual content of these biocultural rights is unclear, and because of the limited resources assigned and the complex institutional situation and other issues that impede the realization of the objectives of Judgment T-622/16 (Vallejo Piedrahíta, McNeish, Moreno Socha, Cuesta Mosquera and Alfonso León this volume).

The Whanganui River

In Aotearoa New Zealand, there is a very explicit recognition of different traditions of knowledge in the Te Awa Tupua (Whanganui River Claim Settlement) Act (2017). Central within the law are the *Tupua te Kawa* – the intrinsic values that represent the essence of Te Awa Tupua. The four *Kawa* are stated in the following way in section 13 of the Act:

- *Ko te Awa te mātāpuna o te ora*: the River is the source of spiritual and physical sustenance
- *E rere kau mai i te Awa nui mai i te Kahui Maunga ki Tangaroa*: the great River flows from the mountains to the sea
- *Ko au te Awa, ko te Awa ko au*: I am the River and the River is me
- *Ngā manga iti, ngā manga nui e honohono kau ana, ka tupu hei Awa Tupua*: the small and large streams that flow into one another form one River

The inclusion of Māori values and laws in this form – very different from the common style of the legal tradition of knowledge of the modern state –

follows from the explicit intention of the Māori negotiators of the settlement to have a law that reflected their knowledge. The fact that the Māori versions of the *kawa* are given pre-eminence in the text emphasizes this. In the political process of agreeing on the settlement, the Māori were unable to achieve ownership of the river, but having their knowledge enshrined in state law was perhaps more important.

Space does not allow a deeper analysis of the *kawa* here. But together they form a holistic vision of the river as an indivisible whole that is intimately connected to and inseparable from the people and communities living along it, to the extent that what is good for the river is good for the people. For the Māori negotiators in the settlement proceedings, this was crucial, as it gave a tool to address what they saw as a major shortcoming of the existing management system: the tendency for "silo thinking", dealing with individual aspects of the river without considering the whole. As Cribb writes:

> The river is viewed and managed separately by the Crown in terms of its bed and banks, its subsoil, its waters, its flora and fauna and its fisheries; the river is subject to a raft of different statutory regimes including the RMA, the Local Government Act, the Fisheries Act, the Conservation Act, the Maritime Transport Act and the Marine and Coastal Area (Takutai Moana) Act; the river comes under the jurisdiction of one regional council, three local authorities and a number of Crown agencies; and issues of navigability, sale and compulsory acquisition mean the bed of the river and its tributaries are in parts privately owned and in other parts in Crown or public ownership.
>
> (Cribb this volume)

The *kawa* of the Te Awa Tupua Act require that decisions related to the river are taken with consideration to the river as a whole. Still, all the legislation Cribb mentions remain in place. But decision-makers are now obliged to apply these laws and regulations in a way that gives regard to the larger picture. This may not always be simple and straightforward and will require politics to have an important place also in the future. But as we have documented in our research: so far, the new law is leading to improved ways of making decisions relating to the river that involve more popular participation and greater attention to the wellbeing of the river (Cribb et al. 2024; Cribb this volume).

Comparing the politics of knowledge

Examining the politics of knowledge in these three cases shows that there are evident parallels in the co-existence of different traditions of knowledge in all cases. Key differences between these co-existing ideas have to do with how to understand and relate to the natural environment. One source of the existence of different sets of knowledge is the three countries' common experience of colonialism, which included the imposition of an imported legal system. One can think of the innovation of legal personhood for the river in all three cases as a way of bridging different knowledge systems, as a form of decolonization.

From such a perspective, the Whanganui case is clearly the most ambitious and the most successful. These cases are a small sample, and one should be wary of generalizing on the basis of them. Yet it does seem very probable that the reason for the Whanganui case being the most promising one is because of the many years spent in negotiation and the participatory way of law-making. This allowed the inclusion of other forms of knowledge into state law. In contrast, the rulings by judges in Colombia and India were essentially top-down decisions that actually in both cases went further than what the claimants had asked for. The decisions were consequently somewhat disconnected from the political context. In contrast, the inclusive – and time-consuming – procedure in Aotearoa New Zealand allowed for deliberation of the details and grounding among stakeholders.

Conclusion

This chapter has analysed how legal personhood for rivers in the three countries must be understood within their political contexts. The aim has been to compare the cases and highlight contrasts as well as parallels. The analysis and comparison have been organized according to three interrelated dimensions: the characteristics of the state, the political economy of resource extraction and the politics of knowledge.

The three states involved are very different in terms of their capacity and degree of sovereignty. While all three countries are former colonies, their colonial history is different. From the perspective of the Māori in the settler colony of Aotearoa New Zealand, it can be argued that colonization never ended. Legal personhood for the Whanganui River can be seen as the partial devolution of sovereignty from the state to the Whanganui Māori and the inhabitants of the river basin, repairing previous injustices and addressing legitimate claims of Indigenous rights. In contrast, in both Colombia and India, the court cases that resulted in legal personhood were initiated as attempts to have the state more forcefully assume its sovereignty in order to remedy the deteriorated environmental state of the rivers.

There are parallels between the cases in terms of the extraction of natural resources being the underlying cause of problems. Yet this similarity should not be exaggerated. The Colombian case is the one where gold mining, poisoning of the river and its people, illegality and organized crime, armed actors and government weakness combine as the interlinked causes of the problem, and this complex state of affairs is also the explanation for the limited impact of the personhood ruling so far. In India, where the rivers face similar contamination issues, extractivism is an important factor among others in causing river degradation and may help explain the limited impact of the ruling. But there are other factors at play, and extractivism is not the evident explanatory key, as is the case in Colombia. In Aotearoa New Zealand, settler colonialism was about taking over the territory and resources of the original inhabitants. Farming and livestock-keeping have had detrimental environmental effects and can be considered a form of extractivism. And while the Te Awa Tupua Act meant that Indigenous claims were addressed, the scope was limited by allowing existing property and use rights to continue. So there is a parallel also here.

In all cases, there are struggles over what forms of knowledge should have precedence. The three decisions of legal personhood for rivers can all be understood as attempts at bridging different forms of knowledge and at promoting ecocentric understandings of the relationship between humans and their natural surroundings. When it comes to actually including local knowledge of a form that is distinct from the knowledge tradition of state law, the Whanganui example is clearly the most successful. It seems reasonable to attribute this success to the participatory and patient negotiation process behind the Te Awa Tupua Act.

Notes

1 The Whanganui River case is described in detail in Cribb (this volume) and in Cribb et al. (2024). For the Atrato River case, see Vallejo Piedrahíta et al. (this volume), McNeish and Moreno Socha (2024) and Cuesta Mosquera and Vallejo Piedrahíta (2024). For the Ganga and Yamuna rivers case, see Borchgrevink and Upadhyay (this volume), Upadhyay and Nayak (2024) and R. Ranjan (2024).

2 Thanks to Anne Waldrop for pointing this out.

References

Alley, Kelly D. 1998. "Idioms of degeneracy: Assessing Ganga's purity and pollution." In *Purifying the earthly body of God: Religion and ecology in Hindu India*, edited by Lance E. Nelson, 297–330. State University of New York Press.

Barth, Fredrik. 1993. *Balinese worlds*. University of Chicago Press.

Barth, Fredrik. 1995. "Other knowledge and other ways of knowing." *Journal of Anthropological Research* 51 (1): 65–68.

Bayley, Nicholas. 2007. *Aspects of the economic history of Whanganui Māori in the Whanganui Inquiry District (Wai 903) 1880–2000*. Waitangi Tribunal.

Benítez-Rojas, Vicente F. 2022. "Beyond invalidation: Unorthodox forms of judicial review of constitutional amendments and constitution-amending case law in Colombia." *Revista de Investigações Constitucionais* 9: 269–300.

Blair, Harry. 2023. "Saving India's rivers: Ecology, civil society, religion, and legal personhood." *World Development Sustainability* 3: 100068.

Borchgrevink, Axel. 2014. *Clean and green: Knowledge and morality in a Philippine farming community*. Ateneo de Manila Press.

Bråten, Eldar. 2016. "Reading Holbraad: Truth and doubt in the context of ontological inquiry." In *Critical anthropological engagements in human alterity and difference*, edited by Bjørn Enge Bertelsen and Synnøve Bendixsen, 273–294. Springer.

Carey, Jane, and Ben Silverstein. 2020. "Thinking with and beyond settler colonial studies: New histories after the postcolonial." *Postcolonial Studies* 23 (1): 1–20.

Chandra, Aparna. 2018. "Courting the people: Public interest litigation in post-emergency India." *International Journal of Constitutional Law* 16 (2) (April): 710–718. doi:10.1093/icon/moy045.

Coatsworth, John H. 2003. "Roots of violence in Colombia: Armed actors and beyond." *ReVista: Harvard Review of Latin America* (Spring): 3–7.

Cribb, Miriama, Elizabeth Macpherson and Axel Borchgrevink. 2024. "Beyond legal personhood for the Whanganui River: Collaboration and pluralism in implementing the Te Awa Tupua Act." *The International Journal of Human Rights*: 1–24. doi:10.1080/13642987.2024.2314532.

Cuesta Mosquera, Arlinton, and Catalina Vallejo Piedrahíta. 2024. "Take me to the River: Have riverine rights enhanced community participation in environmental governance within the Atrato River basin?." *The International Journal of Human Rights*: 1–19. doi:10.1080/13642987.2024.2367468.

DANE (Departamento Administrativo Nacional de Estadística). n.d. *La visibilización estadística de los grupos étnicos Colombianos.*

de Micheaux, Flore Lafaye. 2019. "Politicizations of a sacred river: The story of Gaumukh-Uttarkashi Eco-Sensitive Zone, Uttarakhand, India." *Journal for the Study of Religion, Nature & Culture* 13 (2).

Economic Times. 2018. "940 dams, barrages built on Ganga restricting its flow: Environmentalists." November 24. https://economictimes.indiatimes.com/news/politics-and-nation/940-dams-barrages-built-on-ganga-restricting-its-flow-environmentalists/articleshow/66784449.cms?from=mdr.

Eriksen, Stein Sundstøl. 2010. "The theory of failure and the failure of theory: 'State failure', the idea of the state and the practice of state building." In *Troubled regions and failing states: the clustering and contagion of armed conflict*, edited by KristianBerg Harpviken, 27–50. Bingley.

Finlayson, Chris, and James Christmas. 2021. *He Kupu Taurangi: Treaty settlements and the future of Aotearoa New Zealand*. Huia Publishers.

Geddis, Andrew, and Jacinta Ruru. 2019. "Places as persons: Creating a new framework for Māori-Crown relations." In *The frontiers of public law*, edited by Jason NE Varuhas and Shona Wilson Stark. Hart Publishing.

Graeber, David. 2015. "Radical alterity is just another way of saying 'reality': A reply to Eduardo Viveiros de Castro." *HAU: Journal of Ethnographic Theory* 5 (2): 1–41.

Gray, Vanessa Joan. 2008. "The new research on civil wars: Does it help us understand the Colombian conflict?." *Latin American Politics and Society* 50 (3): 63–91.

Guha, Ranajit. 1989. "Dominance without hegemony and its historiography." In *Subaltern studies* VI, edited by Ranajit Guha. Oxford University Press.

Gupta, Akhil. 1995. "Blurred boundaries: The discourse of corruption, the culture of politics, and the imagined state." *American Ethnologist* 22 (2): 375–402.

Hansen, Thomas Blom. 2019. "The state as an ethnographic object." In *Critical themes in Indian sociology*, edited by Sanjay Srivastava, Yasmeen Arif and Janaki Abraham. SAGE.

Hansen, Thomas Blom, and Finn Stepputat. 2006. "Sovereignty revisited." *Annual Review of Anthropology* 35 (1): 295–315.

Kauffman, Craig M., and Pamela L. Martin. 2021. *The politics of rights of nature: Strategies for building a more sustainable future*. MIT Press.

Khilnani, Sunil. 1998. *The idea of India*. Penguin.

King, Michael. 2003. *The Penguin history of New Zealand*. Penguin Random House New Zealand.

Kumar, Sanjay. 2018. "Indian scientist dies after Gandhi-style hunger strike to save the Ganges River." *Science*, October 16. https://www.science.org/content/article/indian-scientist-dies-after-gandhi-style-hunger-strike-save-ganges-river.

Logan, Owen, and John-Andrew McNeish. 2012. *Flammable societies: Studies on the socio-economics of oil and gas*. Pluto Press.

Maté, Manoj. 2015. "The rise of judicial governance in the Supreme Court of India." *Boston University International Law Journal* 33: 169.

Mawdsley, Emma. 2005. "The abuse of religion and ecology: The Visha Hindu Parishad and Tehri Dam." *Worldviews: Global Religions, Culture, and Ecology* 9 (1): 1–24.

McNeish, John-Andrew. 2021. *Sovereign forces: Everyday challenges to environmental governance in Latin America*. Berghahn Books.

McNeish, John-Andrew, and Judith Shapiro. 2021. "Introduction". In *Our extractive age: Expressions of violence and resistance*, edited by Judith Shapiro and John-Andrew McNeish. Routledge.

McNeish, John-Andrew, Axel Borchgrevink and Owen Logan, eds. 2015. *Contested powers: The politics of energy and development in Latin America*. Bloomsbury Publishing.

McNeish, John-Andrew, and Jenny Moreno Socha. 2024. "Muddy waters: On the problematic political ecology of the Atrato ruling, Colombia." *The International Journal of Human Rights*: 1–21. doi:10.1080/13642987.2024.2324121.

Mikaere, Ani. 2011. *Colonising myths-Maori realities: He rukuruku whakaaro*. Huia Publishers.

Mohd. Salim v. State of Uttarakhand & Others. 2017. March 20 for Writ Petition (PIL) No. 126 of 2014.

Nielsen, Kenneth Bo, and Alf Gunvald Nilsen. 2021. "Hindu nationalist statecraft and Modi's authoritarian populism." In *Routledge handbook of autocratization in South Asia*, edited by Sten Widmalm, 92–100. Routledge.

Olsen, Jonathan Robert. 1999. *Nature and nationalism: Right-wing ecology and the politics of identity in contemporary India*. Macmillan.

Parry, Jonathan. 2000. "The 'crisis of corruption' and 'the idea of India'." In *Morals of legitimacy: Between agency and system*, edited by Italo Pardo, 27–55. Berghahn Books.

Petel, Matthias. 2024. "The illusion of harmony: Power, politics, and distributive implications of rights of nature." *Transnational Environmental Law* 13 (1): 12–34.

Ponnaganti, Amani. 2024. "Toxic formations: Race, place and the politics of pollution on the banks of the Ganga." *International Journal of Urban and Regional Research* 48 (2): 263–279.

Randall, Stephen. 2015. "Nationbuilding and civil war: Diverging views of state and society in late 19th century Colombia." *Journal of Military and Strategic Studies* 16 (3).

Ranjan, Amit. 2024. "Environmental politics in India between 2014 and 2019: A critique of policies and projects." *Social Change* 54 (4): 526–542.

Ranjan, Rahul. 2024. "Sadhu activism: Rights of rivers and facets of religious environmentalism in the Himalayas." *The International Journal of Human Rights*: 1–21. doi:10.1080/13642987.2024.2420244.

Richardson, Whitney. 2020. "Nature's rights in Colombia: An exploration of legal efforts to secure justice for humans and nature." Master's thesis, Norwegian University of Life Sciences.

Richardson, Whitney, and John-Andrew McNeish. 2021. "Granting rights to rivers in Colombia: Significance for ExtrACTIVISM and governance." In *Our extractive age*, edited by Judith Shapiro and John-Andrew McNeish, 155–175. Routledge.

Rotberg, Robert I. 2003. "Failed states, collapsed states, weak states: Causes and indicators." In *State failure and state weakness in a time of terror*, edited by Robert I. Rotberg, 1–25. Brookings Institution Press.

Rudolph, Lloyd I., and Susanne Hoeber Rudolph. 2010. "Federalism as state formation in India: A theory of shared and negotiated sovereignty." *International Political Science Review* 31 (5): 553–572.

Sharma, Mukul. 2009. "Passages from nature to nationalism: Sunderlal Bahuguna and Tehri Dam opposition in Garhwal." *Economic and Political Weekly* 44 (8): 35–42.

Sharma, Mukul. 2023. "Hindu nationalism and right-wing ecology: RSS, Modi and motherland post-2014." *Studies in Indian Politics* 11 (1): 102–117.

Silverstein, Ben. 2016. "Patrick Wolfe (1949–2016)." *History Workshop Journal* 82 (1).

Sundar, Nandini. 2023. "The supreme court in Modi's India." *Journal of Right-Wing Studies* 1 (1).

Tauri, Juan, and Robert Webb. 2011. "The Waitangi Tribunal and the regulation of Māori protest." *New Zealand Sociology* 26: 21–41.

Te Arawhiti (The Office for Māori Crown Relations). 2024. *Treaty Settlements Quarterly Report*, 1 July–31 December.

The Guardian. 2014. "Modi draws huge crowds as election campaign moves to Varanasi." April 24. https://www.theguardian.com/world/2014/apr/24/narendra-modi-election-india-varanasi.

The World Bank. 2014. *Ganges strategic basin assessment: A discussion of regional opportunities and risks world bank South Asia regional report*. The International Bank for Reconstruction and Development / The World Bank.

Tubb, Daniel. 2015. "Muddy decisions: Gold in the Chocó, Colombia." *The Extractive Industries and Society* 2 (4): 722–733.

Upadhyay, Kavita, and Bibhu Prasad Nayak. 2024. "Tracing the legal journey of petitions in the Uttarakhand High Court that became springboards for rights of rivers and nature in India." *The International Journal of Human Rights 0* (*0*): 1–20. doi:10.1080/13642987.2024.2380850.

Veracini, Lorenzo. 2011. "Introducing: Settler colonial studies." *Settler Colonial Studies* 1 (1): 1–12.

Waitangi Tribunal. 1999. *The Whanganui River report*. GP Publications.

Wilson, John. 2007. "New Zealand sovereignty: 1857, 1907, 1947, or 1987?." *Political Science* 60 (2): 41–50.

8 Rights of nature redux

Malene K. Brandshaug and John Andrew McNeish

Introduction

The growing number of cases of granting rights to nature in diverse countries across the planet signals that there is a global move towards eco-centric law-making – from a human-dominated rights regime towards laws that include or centre on nature. Legal cases have been created that deal with all of nature within a country or municipal territory, or specific natural entities, such as rivers, mountain and whole ecosystems. Moreover, there are now many non-governmental organizations advocating for the expansion of rights of nature, often belonging to a larger social movement. What has become a global rights of nature movement is heterogenous, and rights of nature is by no means a unified global phenomenon. As this volume has already demonstrated, each legal case has its own characteristics, shaped by unique local and national circumstances. This said, we highlight in this chapter that there are several global trends that have contributed to opening a space for a heightened interest in eco-centric lawmaking. A central aim of such legal innovations is to combat environmental destruction and ensure the health of ecosystems by recognizing nature's intrinsic value and expanding rights to natural entities.

In this chapter, we consider some of the key global dynamics encouraging the spread and uptake of the rights of rivers specifically, and rights of nature more generally. As such we highlight a series of key dynamics inspiring and directing the current spread of these legal innovations around the world. We observe that in addition to a series of both local and international organizations campaigning in support of the rights of nature, changes in scientific understanding and developments in social theory have also played a significant role in the formation of an expanding zeitgeist in which the once fringe and unthinkable idea – that nature has legal rights – is gaining increasing formal acceptance. In addition to sketching these transformations and demonstrating important and possibly positive alignments, the chapter furthermore indicates that as well as representing opportunities for new insight and legal outcomes for the protection of nature and planet, necessary questions must also be asked about the possible pitfalls of this shift in law.

DOI: 10.4324/9781003581208-8

As we can see from earlier chapters in the book, rulings on the rights of rivers and on the rights of nature have stimulated new ways of thinking about and acting to protect the natural environment. However, we question here whether the shift towards eco-centric law might also blind us to persisting challenges and inequalities and fail to offer a practical mechanism for substantive change. When we conclude with this critical take on rights of nature, the aim is not to disregard the innovative and possibly productive avenues of recognizing the rights of nature, nor the moral shift it represents, but to beware of the possible risks and shortcomings that may come with it.

Rights of nature redux

In "Should trees have standing? Towards legal rights for natural objects" (1972), Christopher Stone proposed granting legal rights to natural entities. Although equating other-than-human entities with humans was not a unique or new thought in a historical and global context, Stone's proposal is often said to be the first to suggest the legal rights of nature in modern law. At the time, this idea was considered "unthinkable" (Stone 1972, 453), despite the long history of legal recognition for other types of other-than-human entities – such as municipalities, corporations and religious institutions – dating back to Roman law in Europe and the common practice of equating other-than-humans with humans in non-modern norm systems. Stone highlighted the modern legal opposition to the rights of natural entities and argued that until an entity receives rights it remains a mere thing for human use (1972, 455). This reflects a modern and capitalist objectification of nature, where nature is viewed as a resource rather than a subject.

Stone's proposed introduction of the rights of nature did not take off in the 70s, but it did inspire further animal rights cases throughout the end of the decade and into the 80s. Moreover, Stone's ideas were developed further in the 80s and 90s (Berry 1988, 1999), and in the early 2000s the broader philosophy of earth jurisprudence (Berry 2001) and wild law (Cullinan 2002) were promoting a new kind of law that places humans within a wider earth community along with other beings. Despite these related developments, Stone's proposal was not taken seriously in the wider legal community or followed by any immediate examples of recognizing the rights of natural entities. More than three decades later, however, the world saw the first practical applications of rights of nature in law, which since have spread and caught on in all continents.

The first instance of modern eco-centric lawmaking occurred in 2006, with a municipal ordinance in Tamaqua Borough, Pennsylvania, although the Navajo Nation acknowledged the rights and freedom of all living entities in line with customary law a few years before (Ecojurisprudence.org 2002). An even more significant breakthrough came in Ecuador in 2008, when the nation's constitution recognized nature (equated in law with the Indigenous Quechua concept Pachamama) as a rights-bearing entity and attempted to incorporate the Quechua Indigenous idea of *sumac kawsay* (often translated as 'living well') – which focuses on relationships of care between humans and nature – as an element of national environmental policy (Constitución de la República de Ecuador 2008). The constitutional change resulted from social mobilization against economic exploitation, Indigenous advocacy and environmentalist influence (Tanasescu 2013). Mihnea Tanasescu argues that while Indigenous resistance played a role in the development of this somewhat radical constitutional change, and especially in the initial movement for change, the advocacy for rights of nature can ultimately be seen as a top-down strategy initiated by a few people in central positions (2013, 855).

In 2011, Ecuador's constitutional rights of nature were first upheld in the Vilcabamba River case, where a court ordered ecological restoration of the river following environmental damage made by the widening of a road.[1]

In 2010 the Bolivian government introduced the Law of the Rights of Mother Earth (Ley de Derechos de la Madre Tierra) (La Asamblea Legislativa Plurinacional 2010), which recognized nature's rights in a broad sense and required a change to take place in the country's economic thinking. However, the actions of the governments in both countries have contradicted their radical intent since their introduction. More recent court cases have attempted to strengthen the position of rights of nature in Ecuador (Tanasescu et al. 2024, 4).

A related case of granting rights to nature occurred in 2014, when Aotearoa New Zealand granted legal personhood to the Te Urewera national park (Te Urewa Act, Department of Conservation 2014), before the better known case of legal personhood to the Whanganui River was recognized through the Te Awa Tupua (Whanganui River Claims Settlement) Act in 2017. Similar cases followed, including Colombia's Atrato River (Sentencia T-622, Corte Constitucional de Colombia 2016) and India's Ganga (Ganges) and Yamuna Rivers (Mohd. Salim v. State of Uttarakhand Case 2017). These were the first cases of giving rights to specific natural entities and not the environment as a whole – as was the case with Ecuador and Bolivia. It is evident from our research on these specific examples – see previous chapters – that legal frameworks vary, but each contribute to a broader trend in eco-jurisprudence.

There are currently about 500 recognized eco-jurisprudence initiatives, many of which take the form of rights of nature. These stretch across six continents, with the highest concentration in the Americas.[2] And even more initiatives are dawning. In Europe, Spain granted rights to the Mar Menor lagoon in 2021, and other proposals are emerging in, among other places, the United Kingdom, Netherlands and France, and the European Union has drawn its attention towards rights of nature. Rights of nature initiatives manifest at multiple legal levels, from municipal ordinances to constitutional law and international institutions, influenced by grassroots activism, court rulings and international advocacy. Some result from long-term, locally specific processes; others are inspired by similar cases elsewhere, larger political-intellectual developments or the work of international organizations advocating for the rights of nature. Often a combination of several factors shapes each case (Vallejo Piedrahíta, McNeish, Moreno Socha, Cuesta Mosquera and Alfonso León this volume, Cribb this volume, Borchgrevink and Upadhyay this volume).

The Ecuadorian and Aotearoa New Zealand cases have become key reference points for rights of nature developments worldwide. These cases are said to represent two different ways of granting rights to nature, either to nature as a whole or delimited to specific natural entities or ecosystems, denominated "the nature's rights model" and "the legal personhood model" in their ideal forms (Kauffman and Martin 2021). Both cases involve Indigenous concerns. Many subsequent rights of nature initiatives explicitly reference these precedents, where the Ecuadorian case is seen as the pioneer case and the Aotearoa New Zealand Whanganui River case as one of particular interest for those concerning rivers and water bodies, which form a significant portion of current rights of nature cases.[3] Apart from being inspired by other cases, rights of nature laws and contemporary rights of nature debates draw from Indigenous ontologies emphasizing reciprocal relationships with nature and from Euro-American environmental philosophy, such as Arne Næss's "deep ecology" (1973). Commonly included in stories about the inspirations behind rights of nature are also references to Stone's (1972) work and to earth jurisprudence, influenced by thinkers like Thomas Berry (2001) and Cormac Cullinan (2002), who advocate for fundamental rights for all beings and critique anthropocentric legal systems. Rights of nature is often framed as a fusion of these intellectual traditions (see among others García Ruales et al. 2024, Tanasescu 2022, Brandshaug and Borchgrevink this volume).

The *redux* mentioned in the chapter title refers to a return but also the reworking and expansion of the practical application of rights of nature within modern, liberal law that up until recently only treated nature as an object for human use. The rapid expansion of rights of nature from a hypothetical idea that was "unthinkable" to a global legal phenomenon stems from shifting environmental concerns, increased legal precedents and growing awareness of ecological crises. The recognition of nature's rights is now one of the fastest-growing legal movements, with David Boyd (2017) – former UN Special Rapporteur on Human Rights and the Environment – claiming that it is "a legal revolution that could save the world". As more legal systems incorporate rights of nature principles, the idea is becoming increasingly mainstream, challenging dominant legal structures and redefining how human-nature relationships are treated within them. In the following sections, we consider a series of wider shifts that are reinforcing the mainstream acceptance of rights of nature.

Global institutions and movements for the rights of nature

Whereas legal rulings on the rights of nature can appear to be the result of domestic legal and political processes, it is important to consider their occurrence within the much larger global landscape of activism, political campaigning and eco-centric lawmaking. In particular, the work of the new Centre for Democratic and Environmental Rights (CELDF) has been influential in advocating for expanding rights beyond humans and for drafting the first rights of nature legislations (Tanasescu 2022, 53–54, 144). Several international campaigns now exist promoting the idea of the rights of rivers and of nature. These include the work of the Earth Law Centre, founded in 2008, who describe their own mission as being to "advance eco-centric laws, policies, and governance for the well-being of the Earth community"[4]; the Global Alliance for the Rights of Nature (GARN), "a global network of organizations and individuals committed to the universal adoption and implementation of legal systems that recognize, respect and enforce Rights of Nature"[5]; and the International Rights of Nature Tribunal, founded in 2014, "to create a forum for people from all around the world to speak on behalf of nature, to protest the destruction of the Earth – destruction that is often sanctioned by governments and corporations – and to make recommendations about Earth's protection and restoration".[6] Additionally, the World Commission of Environmental Law (WCEL) – that works to strengthen environmental law and policy across the world – has a Rights of Nature Task Force that seeks to "initiate a strategy for dissemination, communication and advocacy on rights of nature".[7]

The international non-governmental organization International Rivers campaigns to protect rivers and defend the rights of communities that depend on them.[8] International Rivers, together with others, promote a Universal Declaration of the Rights of Rivers, urging governments to implement rights of rivers and to "stand with a growing international movement to protect rivers, freshwater ecosystems, and all who depend upon them for their lives and livelihoods"[9]; GARN Europe runs a campaign for the rights of aquatic ecosystems[10]; Embassy of the North Sea aims "to represent the plants, animals, microbes, and people in and around the North Sea in the cultural, legal and political public domain"[11]; and the related Confluence of European Water Bodies advocates political representation for water bodies to "demand attention for the many crises of water today".[12] It is also important to recognize that there is a cross-fertilization between these non-governmental campaigns and others that are dedicated to the recognition of animal and species rights.[13] Further connections can also be made to social movements for water justice, environmental justice and climate justice, as well as Indigenous movements.

The rights of nature and rivers forms part of a "new water justice movement" (Boelens et al. 2022) that incorporates other-than-human matters in "a colourful assembly of grassroots groups and initiatives, as well as regional networks and non-governmental alliances, that mobilize to protect or revive rivers, and to challenge dominant ways of understanding, ordering and exploiting rivers and riverine inhabitants" (Hommes et al. 2023, 1). Also, given the rights of nature movement's connection to Indigenous concerns, it continues a history of heterogenous alliances between environmental movements and Indigenous movements. Indigenous people are increasingly voicing their environmental concerns related to resource exploitation and contamination or theft of their waters (Yazzy and Risling Baldy 2018). In Aotearoa New Zealand, the granting of personhood rights to Whanganui River is a result of a long-lasting and ongoing Indigenous fight over land and water. In India, Ecuador and Colombia, the granting of rights to nature was a top-down decision. However, Indigenous and other grassroots protests were behind a larger call for change addressing socio-environmental injustices. The Latin American cases can also be tied to a larger Indigenous political mobilization in Latin America, where it is argued by some scholars that "earth-beings" have entered politics (de la Cadena 2010). Hence, although not behind the specific legal tools and decisions, Indigenous movements have played a significant role in the broader picture of influences towards rights of nature.

Concurrent with these social movements and non-governmental efforts, governments and international governmental organizations have also made a series of relevant actions contributing to a changing understanding of the place of the environment in rights and law. Reflecting the broader tendency for sustainable development over the last three decades, the provisions for environmental protection in national constitutions have expanded dramatically. It is estimated that 150 constitutions have some reference to environmental protection (O'Gorman 2017). While the United Nations (UN) Conference on the Human Environment in Stockholm in 1972 is rightly credited with initiating the drive towards sustainable development and environmental constitutionalism (Boyd 2011), most environmental policy changes have taken place since the 1990s (May 2006). Many national constitutions in Latin America were rewritten in the 1990s, reflecting an era in which dictatorships ended and rights protections were expanded to previously marginalized sectors of the population such as Indigenous peoples, women and children. There was also increasing awareness about climate change and environmental degradation spurred by social movements concerned with issues such as deforestation in the Amazon, or industrial and infrastructural development in Europe. The growth in normative environmental provisions was accompanied by a growing judicialization of politics, in which the domain of the litigator and judge was radically expanded, and activist lawyers and judges were increasingly the producers of legal reform, especially in constitutional law and the field of rights (Shapiro and Stone Sweet 2002, 19).

The linking of environmental interest to existing procedural and substantive human rights protections has assisted the project of global environmental constitutionalism, as in most legal systems human rights already have constitutional status (Shelton 2020). The connection between human rights and environmental rights was already evident in the declaration made following the 1972 UN Conference on the Human Environment. Following this, many countries have included "the right to a clean and healthy environment" in their national constitution (Shelton 2020, 163) and have been referred to as *Fundamental Environmental Rights* (May 2006). More recently, the UN General Assembly introduced a resolution in July 2022 that everyone on the planet has the right to a clean and healthy environment.[14] This is now recognized by 80% of UN Member States.[15] It is also of note that parallel to the UN's push for recognition of the right to a healthy environment it has also been exploring the significance of the rights of nature directly. In 2015, the UN General Assembly called for the creation of an expert report on Earth Jurisprudence, establishing a global network of experts. The Expert Report on Earth Jurisprudence that followed was released in 2016.[16] The Report recognizes the "fundamental legal rights of ecosystems and species to exist, thrive and regenerate". In 2017, the UN General Assembly Dialogue focused on applying Earth Jurisprudence to the Sustainable Development Goals.[17] Following this, the UN General Assembly adopted in December 2024 the Harmony with Nature resolution that reaffirms their commitment to several environmental declarations and agreements, recognizes the existence of rights of nature and acknowledges the multiple ways of conceptualizing Nature or Mother Earth – reflecting the interdependency between humans, other species and the planet. This initiative promotes a holistic approach to sustainable development that ensures harmonious living with nature.[18] In Europe, the EU has also drawn its attention towards rights of nature with the two reports "Towards an EU Charter of the Fundamental Rights of Nature" (2020)[19] and "Can Nature Get It Right?" (2021).[20]

This increased international interest for earth jurisprudence is part of a larger move towards progressive environmental laws. For instance, the International Court of Criminal Justice has been urged to criminalize ecocide to hold powerful actors accountable for their environmental destruction through international law.[21] Moreover, the mentioned developments in international environmental law have led to a growing number of climate litigation cases globally,[22] many of which are directed towards governments and companies responsible for negative climate impacts.[23]

It is important to acknowledge that these positive tendencies have not been without controversy, being in part responsible for a backlash against environmental policy. This is clearly linked to the rise of right-wing populist politics in Europe, the Trump Presidency retraction of the US from the Paris Agreement and new global power imbalances in the wake of increased international conflict and uncertainty. In the context of rights of nature, Aotearoa New Zealand is now risking a backlash against further treaty settlements with Māori, due to the election of a government with less progressive environmental and Indigenous politics. Although Colombia continues to have a progressive government that is interested in strong environmental policy, with the recent turn to the political right in many Latin American countries there has been a return to a more conservative understanding of rights and land use and the further expansion of resource extractive activities more broadly in the region. Despite the enshrinement of Hindu religious values venerating the divine relationship between the gods and the environment within its political platform, the Modi government has also acted to extensively expand on resource extractive activities such as mining, which has direct consequence for the health of rivers, people and the nation's wider biodiversity.

The recent increase in the number of rights of nature cases worldwide is as highlighted above driven by a broad constellation of movements, organizations and ideas (not without opposition). In the next section we sketch some of the broader trends in modern science and the development of new understandings of humans' place in the world.

Our place in the nature world

Recent discoveries across various natural sciences – ecology, botany, ethology, mycology, microbiology, geology, chemistry and others – are increasingly challenging the long-standing distinction between humans and other-than-humans. This shift in understanding not only disrupts the anthropocentric perspective that has dominated fields such as human rights but also necessitates a re-evaluation of the human-nature relationship. Scientists, equipped with advanced tools such as artificial intelligence and biosensors, are now eavesdropping on the languages of nonhuman species, attempting to decipher whale calls, bird songs and even the communication methods of mole rats (Crance at al. 2019, Taylor and Cody 2019). These investigations suggest that linguistic complexity and social structures extend far beyond the human domain. Meanwhile, botanists are tracing underground mycelial networks that enable trees to exchange nutrients and chemical signals in what has been termed the "wood wide web" (Simard et al. 2015), and mycologists are mapping fungal ecosystems, revealing a vast, hidden biosphere that shapes terrestrial life in profound ways (Sheldrake 2021).

The deeper scientists explore these entanglements, the blurrier the boundary between individuals and ecosystems becomes. Human bodies, often considered distinct biological entities, are in fact complex ecological assemblages of microbial communities. The gut microbiome, for example, plays a crucial role in digestion, immunity and even neurological processes, further demonstrating the inseparability of humans from their microbial counterparts (Gilbert et al. 2012). Merlin Sheldrake (2021) emphasizes that symbiosis is not an anomaly but a ubiquitous feature of life, underscoring the interconnectedness of all living systems.

This growing recognition of interdependence resonates with earlier ecological theories, such as James Lovelock's Gaia hypothesis (Lovelock 1979). Lovelock proposed that the Earth operates as a self-regulating system, wherein biological and geochemical processes interact to maintain conditions conducive to life. His ideas, though initially controversial, have gained traction in contemporary research into planetary boundaries and Earth system science. The Anthropocene concept, introduced to capture the profound impact of human activity on planetary processes, further complicates this picture. As Will Steffen et al. (2015) argue, industrialization, deforestation and greenhouse gas emissions have pushed the Earth beyond its safe operating limits, endangering biodiversity, climate stability and oceanic systems. Johan Rockström et al. (2009) outline key planetary boundaries that humanity must respect to avoid catastrophic shifts, including climate change, biosphere integrity and biochemical flows. The recognition that humans are altering fundamental Earth processes necessitates a radical rethinking of our place within nature, requiring transdisciplinary collaborations across natural and social sciences (Lenton et al. 2019).

Beyond Earth, scientific discoveries continue to reshape our understanding of humanity's place in the universe. Since ancient times, the study of celestial bodies has been central to human thought, influencing agricultural practices, navigation and cosmological beliefs. The space age, beginning in the late 1950s, has further transformed these perceptions by providing new vantage points on Earth and the cosmos. The famous "Earthrise" photograph, taken during the Apollo 8 mission, catalysed environmental movements by revealing Earth's fragility against the vastness of space (Cosgrove 1994, Dunnett et al. 2019). Today, satellite imagery reinforces these concerns, offering unprecedented insights into deforestation, ice sheet melting and urban expansion, thereby strengthening calls for sustainability.

The contemporary "new space" era, characterized by a surge in private sector involvement and ambitious extraterrestrial colonization plans, raises additional ethical and ecological considerations. Companies such as SpaceX, Blue Origin and national space agencies are not only increasing the frequency of rocket launches but also pursuing long-term settlement goals on the Moon and Mars. These ventures herald a new phase of exploration, but they also pose significant risks, including space debris accumulation, planetary contamination and unforeseen environmental consequences beyond Earth. Importantly, these developments concurrently also challenge conventional understandings of "the environment" by expanding its conceptual and geographical scope beyond Earth. Through her ethnographic study of planetary scientists, astrobiologists and space explorers, Lisa Messeri (2015) demonstrates how scientific and technological engagements with Mars, exoplanets and other celestial bodies reshape not only our physical explorations but also our intellectual and cultural understandings of environments and habitability. Traditionally, "the environment" has been conceived as the Earth's biosphere, encompassing forests, oceans, atmosphere and the intricate web of terrestrial and aquatic ecosystems. Messeri (2015) highlights how planetary scientists are constructing new understandings of environments on other planets by defining habitability in terms of water presence, atmospheric composition and geochemical cycles. This shift encourages us to consider environments beyond biological ecosystems, incorporating geophysical and atmospheric conditions that may be hospitable to life in alternative forms.

As we push deeper into space, questions arise about the nature of sustainability in an interplanetary context. There is a danger that world settlements will replicate the exploitative dynamics that have led to ecological crises on Earth. The principles of planetary stewardship now need to extend beyond our home planet. Ultimately, the shifting scientific understanding of interspecies entanglements, planetary boundaries and extraterrestrial environments calls for a profound re-evaluation of human identity. As we increasingly recognize that we are not isolated beings but participants in a complex, symbiotic web of life, ethical and political frameworks must evolve accordingly. This means moving beyond traditional human-centric paradigms and fostering an integrated approach that considers the rights, agency and intrinsic value of all life forms – both on Earth and beyond.

The end of Enlightenment?

In *We have never been modern*, Bruno Latour (1993) criticizes the notion of modernity and explores what he identifies as distinctive modernist practices. He examines its foundational dualism, particularly the separation of humans from nonhumans through "purification practices". At the same time, he highlights "hybridization" or "translation", wherein nature and culture intermingle, even within scientific practice. It is the consciousness of and confusion of these sets of dichotomous practices that lead to a realization that we have never been modern (1993, 20–21). As such, Latour's analysis challenges the modernist assumption of a strict nature-culture divide and highlights how this idea continues to be reproduced by modern institutions. Urging recognition of the interwoven relationships between modern dualisms, his work has significantly influenced social theory, particularly in challenging Enlightenment-based scientific thinking and Cartesian dualism.

The Cartesian divide, established by René Descartes in the 17th century, introduced a rigid separation between mind and body, a dualism that underpins much of modern Western philosophy. This binary logic extends beyond mind-body distinctions to broader categories such as human-nature, subject-object and past-present, shaping dominant worldviews and legal structures. Modernity, often characterized by its universalist aspirations, assumes these distinctions as natural and applicable everywhere. However, for the past few decades, scholars across the social sciences and humanities have challenged these binaries, arguing that lived experience seldom adheres to such strict separations (Latour 1993). Anthropologists, for instance, have demonstrated through ethnographic research that human and nonhuman entities are deeply entangled in daily life, leading to what has been termed the "more-than-human" turn.

The more-than-human turn – which extends beyond anthropology to social sciences and humanities more broadly, to fields such as philosophy, literature, arts and, as the movement for rights of nature testifies to, law – seeks to move beyond entrenched human exceptionalism, emphasizing interrelations between humans and other-than-humans. Although encompassing somewhat diverse currents, there is a common curiosity towards how specific more-than-human entities matter for one another regardless of dominant dualistic categorizations – how animals, beings and things shape human lives and vice versa. Drawing on and exceeding the more established conceptual framework of political ecology and environmental anthropology, this trend consists of several related streams and inspirations, such as multispecies ethnography and Science and Technology Studies (STS), as well as broader theoretical movements like posthumanism, neo-materialism and decolonial thought.

Critical posthumanism interrogates human exceptionality and supremacy by challenging the dominant figure of the universal human subject – typically conceptualized as the "average adult-white-heterosexual-European-male-speaking a standard language" (Deleuze and Guattari 1987, 105). This resonates with the critical stance of feminist, queer, black and postcolonial theory. Central scholarship within these turns opposes anthropocentrism, emphasizes planetary interconnections and looks for new ways to understand complex human-environment relationships. This may entail rethinking ideas of sociality and agency beyond the human (Latour 1993, Tsing 2013, Kohn 2013), acknowledging other-than-humans as more than passive objects (Barad 2003, Bennett 2010), drawing attention to how humans dwell and are embedded in the environment (Ingold 2000) and how we "become *with*" other species (Haraway 2008, 77), and studying the social and material together, focusing on how the world comes into being through relational practices (Law 2004, Barad 2003, Mol 2002, Latour 1993).

A wider interest in ontological processes expands these critiques by arguing for ontological multiplicity (Mol 2002) and how the world may consist of multiple ontologies, realities or worlds (de la Cadena and Blaser 2018, Descola 2013, Viveiros de Castro 2004, Escobar 2020). This challenges "the one-world world" (Law 2015), the modern ontological dominance (Blaser 2013), and entails a politics concerned with negotiations over what there is, variously termed ontological politics, cosmopolitics or pluriversal politics (see Stengers 2010 [1997], Latour 2004, de la Cadena 2010, Escobar 2020, Blaser 2013). As such, political ontology challenges an all-encompassing modernity and is open for the possibility to understand radical difference in terms of a difference in ontology. These latter concerns are shared by central rights of nature scholarship on onto-legal pluralism, which is inspired by the more equal status and interrelatedness of humans and other-than-humans within Indigenous "relational ontologies" (see Brandshaug et al. this volume).

While a radical potential for better environmental management can be found within Indigenous relational ontologies (Tanasescu 2022), some scholars warn that drawing inspiration from Indigenous being and thinking is not unproblematic. Important scholarship on Indigenous methodology and decoloniality draws attention to extractive research and knowledge retraction from Indigenous peoples in (post)colonial settings. This scholarship warns against appropriation and essentialization of that which is assumed to be "purely"

Indigenous. For instance, the Indigenous Māori scholar Linda Tuhiwai Smith (1999) thematizes the connection between research, colonialism and experiences of exploitation, drawing on the Aotearoa New Zealand history and context. Similarly, Ngugi wa Thiong'o (1986) points to the "colonisation of the mind" in an African setting. In Latin America, the work of Aníbal Quijano (2000) and Walter Mignolo (2007) on the "coloniality of power" and the "de-colonization of knowledge" have inspired a larger body of decolonial scholarship in the region that further problematizes modernity's continued epistemological dominance, emphasizing the persistence of colonial power structures in contemporary knowledge production.

In challenging modernity, hierarchies and binaries, they are not only confronting the ongoing coloniality of Indigenous people but also of nature – which resonates with central rights of nature discourse. As mentioned, inspiration is drawn from Indigenous holistic and often animate ways of conceiving of and relating to the environment, but also from natural sciences such as biology, ecology and even planetary system thinking, which make visible connections and flows in the environment of which humans are part. The Gaia theory (Lovelock 1979), for instance, named after the Greek goddess as a picture for the living planet, has inspired much social theory thinking. The philosopher Isabelle Stengers (2010 [1997]) both appeals to and reimagines Gaia theory as developed by the chemist James Lovelock and the biologist Lynn Margulis, in a feminist project of exploring multiple lifeforms outside of human exceptionalism and capitalist domination.

> Invoking Gaia is a gesture toward a feminist politics of the earth, one that reorients attention from the Anthropos of the Anthropocene, a figure that belies the false universality of Man, to the many earthly forces that partake in world-making projects.
>
> (Tola 2016, 17)

Hence, posthuman, feminist and decolonial critiques resonate within contemporary environmental discourse, particularly in response to the Anthropocene. Emerging as a concept aimed at describing the central characteristics of the current planetary epoch following the Holocene, where *Anthropos*, the Greek word for human, has altered and contributed to permanently changing the earth's climate, geology and ecology (Steffen et al. 2007), the concept has been much debated within social theory. On the one hand, the Anthropocene is disruptive of the nature-culture divide by attracting attention to how humans have impacted nature in pervasive ways, showing that these categories cannot be understood in separation from each other. On the other, the dominant narrative of the Anthropocene has been critiqued for not differentiating the story and putting a homogenous human species as the central actor in the world (see among others Chakrabarty 2019, Haraway 2016, Swanson et al. 2017). Scholars have instead suggested alternative terms, such as the Capitalocene – emphasizing capitalist activities rather than human activities in general as the central driver behind the extensive environmental

Figure 8.1. Wall painting in Quibdó, Colombia. Photo by Daniel Alfonso León.

impacts – (Malm and Hornborg 2014, Moore 2016) and the Chthulucene – drawing attention to human-nonhuman inseparability – (Haraway 2016); and advocated for localized narratives of Anthropocene landscapes that tell differentiated stories of human-nonhuman entanglements (Mathews 2017, Swanson et al. 2017, Tsing 2015).

These theoretical developments in social theory, with an increased recognition of the intrinsic connection between humans and the environment that challenges dualistic, universalistic Enlightenment thought, have given space for the agency of other-than-humans, in many instances inspired by Indigenous lifeworlds. Through the rights of nature movement, these currents have influenced the legal recognition of nature as a subject of rights. However, although the more-than-human legal turn challenges the historical treatment of nature as an object for human use, the institution of liberal law – including the concepts of rights and legal personhood – deeply relies on the modern divide between person and thing, subject and object, culture and nature (Anderson 2024, 53). Hence, while rights of nature represents a significant shift in legal thinking, it also risks reproducing the binaries it seeks to dismantle, inadvertently reinforcing anthropocentric legal structures. Moreover, the movement's reliance on Indigenous ontologies, while valuable, raises concerns about epistemic appropriation and the essentialization of Indigenous knowledge.

Ultimately, the move towards more-than-human lawmaking signals the prospect of a profound shift away from Enlightenment thought, offering new possibilities for reimagining human-nature relations within legal and political frameworks. The idea also stipulates a potential for decolonialization of people and nature. However, the rights of nature movement must remain attentive to its own limitations, ensuring that it does not replicate the exclusions and hierarchies it aims to transcend. Without addressing these blind spots, the potential for a truly transformative paradigm shift may be undermined.

Potential risks and pitfalls

While this chapter has explored the many productive avenues of thought and practice that can be said to inspire as well as work in concert with the rights of nature movement, it would be remiss not to acknowledge the potential pitfalls that accompany these developments. Several key points of friction and unintended consequences warrant careful consideration, as they may compromise the movement's progressive aspirations and, in some cases, risk reinforcing the very systems it seeks to dismantle. When addressing the risks associated with rights of nature, we recognize that these critiques are not valid for all rights of nature frameworks. The three riverine rights cases discussed in this book, for instance, are to different degrees subject to the points we make. Yet, there are general concerns related to the global rights of nature movement that should be addressed. What is presented here are risks and pitfalls that are true for some threads of this heterogeneous eco-jurisprudence movement. More than the idea of granting rights to nature, what calls for critical attention is how this idea is used in rights of nature advocacy and legislation, and what unintended consequences might arise. Critical understandings and self-reflection are needed for rights of nature initiatives to be productive and transformative in addressing socio-environmental issues (see also Boelens et al. 2024).

The rights of nature movement is not merely a legal innovation aimed at expanding rights and enhancing environmental protection; it also functions as a critique of modernity itself, challenging entrenched anthropocentric legal frameworks and epistemologies. Some proponents of rights of nature, particularly those influenced by Indigenous and decolonial perspectives, propose a radical rejection of European conceptualizations of modernity. This critique is essential for addressing modernity's colonial legacies, its environmental destructiveness, Eurocentrism and the nature-culture binary. However, an outright dismissal of its foundational principles risks undermining the very frameworks that have enabled scientific knowledge production and the development of universal human rights. Scientific consensus is necessary to address pressing global challenges, such as climate change, public health and technological innovation. Some postmodern and decolonial critiques, though valuable in interrogating power structures in knowledge production, have been appropriated by anti-science movements to contest well-established scientific truths. Hence, it is necessary for rights of nature to carefully differentiate between the oppressive structures of modernity and its epistemological achievements, in order not to inadvertently weaken scientific literacy and critical engagement with global crises.

Beyond its epistemological dimensions, the rights of nature also presents potential risks in the realm of social justice. Without sufficient attention to equity and democratic participation, the movement risks obscuring structural exploitation and reinforcing historical injustices rather than promoting equality and empowerment. The rise of eco-centric lawmaking, which grants legal rights to nature, has been hailed as a progressive step towards addressing environmental degradation. However, if applied without careful contextual analysis, such legal frameworks may obscure persisting inequalities related to race, ethicity, class and gender. By centring nature as a rights-bearing entity, eco-centric legal mechanisms may overlook the disproportionate environmental burdens faced by marginalized communities, such as Indigenous peoples, low-income groups and communities of colour.

Environmental justice scholars have long argued that those who contribute the least to environmental destruction often suffer its worst consequences (Bullard 1993). For instance, strict conservation laws have historically led to the displacement of Indigenous communities from their ancestral lands, reinforcing colonial power structures rather than dismantling them. Such social dimensions are differently addressed in current practical examples of rights of nature. Here one can distinguish between rights of nature frameworks that purely place nature as the rights-bearing instance – for instance, the rivers Ganga and Yamuna in India – and those that to varying degrees of success place people and nature together as the relevant rights holder. The example from Colombia of recognizing the biocultural rights of Atrato River and its local communities does so, at least in theory. The case of Whanganui River in Aotearoa New Zealand also accomplishes this in practice, to a large extent. In the case of rights of nature in Ecuador with great emphasis on Indigenous injustices, there are examples of rights of nature being used both to the advantage and disadvantage of local people (Boelens et al. 2024). This underscores the need for rights of nature to thoroughly and explicitly address these social dimensions, to avoid perpetuating historical injustices rather than remedying them.

Eco-centric legal reforms may also deepen gender and class inequalities. Women, particularly in the Global South, often bear primary responsibility for securing water, fuel and food for their households, making them especially vulnerable to restrictive environmental policies (Agarwal 1992). Policies that prioritize ecosystem protection over human livelihoods, which is true for some strands of eco-centric lawmaking, can also further marginalize working-class communities who rely on natural resources for subsistence and economic survival. Additionally, the governance structures overseeing eco-centric laws are often shaped by elite policymakers and legal scholars, who may be disconnected from the lived realities of those most affected by environmental and social injustices (Shiva 1989). Moreover, de-centring humans may lead to "overlooking the lived realities of riverine families, livelihoods, meanings, values, identities, and rights systems" (Boelens et al. 2024, 658).

Our research shows that the degree of success in addressing local environmental and human concerns together is linked to the extent of participatory processes involving a representative selection of the local population in river management (Borchgrevink and Brandshaug this volume). Without mechanisms for democratic participation and local engagement, rights of nature laws risk becoming instruments of exclusion rather than empowerment.

Furthermore, while the expansion of nature's rights within legal systems represents a significant departure from anthropocentric legal traditions, it does not necessarily challenge the underlying economic structures that drive environmental harm. Corporations and governments can co-opt eco-centric laws to engage in greenwashing while continuing exploitative practices, masking ongoing inequalities rather than addressing them (Dunlap and Sullivan 2020). In Ecuador, for instance, there are examples of rights of nature being used by government and commercial actors to legitimize eco-business exploitation and eviction of local people from their territory (Boelens et al. 2024, 657). Hence, without careful consideration of its embeddedness in power hierarchies and larger political-economic structures, the ecologically oriented and liberatory potential within rights of nature can be abused by powerful actors.

Despite its theoretical alignment with decolonial and post-humanist thought, the shift towards eco-centric lawmaking often reproduces the very binaries and essentialist distinctions it seeks to dismantle. While more-than-human perspectives aim to move beyond the human-nature dualism, eco-centric legal frameworks frequently reinforce this separation by granting legal personhood to ecosystems while maintaining anthropocentric structures of governance (Plumwood 1993). Similarly, although decolonial discourse emphasizes Indigenous cosmologies and relational ontologies, many rights of nature laws translate these lifeworlds into static legal categories that risk essentializing Indigenous knowledge and reducing it to a symbolic function within Western legal traditions (Todd 2016). This process echoes colonial epistemological frameworks, which historically appropriated Indigenous ways of knowing while simultaneously subordinating them to Eurocentric legal and political systems (Maldonado-Torres 2007). It is worth noting that many Indigenous people also strategically use these legal and political systems as well as their own onto-epistemologies to address socio-environmental issues. Without fundamentally altering the structures that sustain environmental and social injustices, rights of nature may ultimately reinforce the very hierarchies and binaries it purports to challenge.

The rights of nature movement represents an attempt to move beyond the limitations of modernity, but its effectiveness depends on how it navigates these complexities. If the rights of nature is framed purely as a rejection of modernity, it risks losing legal credibility and practical enforceability. If, however, it is seen as a productive hybrid of ecological and Indigenous perspectives with scientific knowledge and human rights protections, it could serve as a powerful legal innovation for socio-ecological justice. Rather than rejecting modernity outright, rights of nature could contribute to reconfiguring modernity – integrating relational ontologies, ecological science and alternative legal philosophies while maintaining the core principles of justice and accountability. This path might allow for a pluralistic yet effective legal framework that acknowledges both the rights of nature and the rights of vulnerable human communities.

Concluding remarks

This chapter has explored some of the key global dynamics that contribute to a renewed focus on the rights of nature. We acknowledge here that while local processes and circumstances shape each case, the phenomenon is characterized by its ability to travel, transform and take on hybrid forms in diverse contexts. Although "rights of nature" does not represent a singular, monolithic legal concept, the overarching trend towards eco-jurisprudence is evident on a transnational scale.

This chapter has sought to understand this trend in relation to several contemporary developments. One such factor is the role of international institutions and organizations that advocate for the rights of nature, which must be viewed within the broader history of environmental, Indigenous and human rights movements. Additionally, evolving scientific understandings of humanity's place in the natural world have emphasized the deep interdependence between humans and ecosystems, which ought to be reflected in new ethical, political and legal frameworks. These scientific insights align with a growing shift in social theory towards recognizing more-than-human interrelations, challenging the primacy of human exceptionalism, modern dualisms and Enlightenment thought. Beyond identifying these driving forces, this chapter has also examined both the opportunities presented by this innovative shift in legal thinking and the potential risks associated with its implementation.

There is little doubt that the idea of recognizing rights for nature has gained momentum and widespread appeal. This is largely due to its fundamental challenge to dominant legal and philosophical traditions that have positioned humans as the masters of nature, a perspective that has contributed to extensive ecological destruction. The current dominant legal frameworks, along with prevailing environmental policies, have repeatedly failed to adequately protect natural ecosystems and the communities that depend on them. When the status quo is demonstrably ineffective – both for the environment and for human wellbeing – the legal recognition of the interconnectedness between humans and other-than-human entities, as well as the extension of legal standing to nature, appears both timely and necessary.

Given the urgent need for alternative approaches, some argue that recognizing rights for nature is, at the very least, a step in the right direction. Jenny García Ruales et al. (2024) contend that within legal frameworks that do not recognize the rights of nature, nature will always be at a disadvantage vis-a-vis human and societal priorities. However, it is equally important to recognize that granting legal rights to nature does not automatically prevent environmentally harmful activities. The transition from a well-intentioned legal principle to effective legal instruments and enforceable practices is complex, and there is no straightforward path from conceptualizing rights of nature to achieving meaningful environmental protection. The diversity of legal interpretations and outcomes observed across various cases demonstrates that recognizing nature's rights does not necessarily guarantee improved conditions for ecosystems. The riverine rights cases examined in this book illustrate this point, highlighting how legal recognition alone does not inherently secure the wellbeing of the ecosystems in question (Borchgrevink and Brandshaug this volume).

Beyond the challenges of implementation, there are broader concerns about whether the rights of nature can truly address the root causes of environmental harm. Just as human rights, in theory, apply universally but in practice are unevenly distributed along lines of ethnicity, class and gender (Douzinas 2000, in Tanasescu 2022, 11), the rights of nature may also be subject to similar limitations. Tanasescu (2022, 14) argues that legal rights are often co-opted to serve capitalist interests and facilitate global economic expansion. This raises questions about whether the rights of nature can effectively counteract the very forces that drive ecological destruction and function as a true countermeasure to the environmental crises engendered by global capitalism. Additionally, the assumed alignment between Indigenous philosophies and the legal recognition of nature's rights is more complicated than it may appear (Tanasescu 2022, 20). Despite the way the rights of nature stipulates decolonialization of vernacular people and nature, the risk remains that rights of nature frameworks may essentialize or appropriate Indigenous worldviews within modern legal structures, reinforcing rather than dismantling existing power hierarchies.

Although law can serve as a starting point for a broader paradigm shift, it must be acknowledged that modern liberal legal systems are deeply embedded in larger political and economic structures. The risks and limitations of the rights of nature must be carefully considered, particularly given the challenges that have historically hindered existing environmental laws from achieving meaningful outcomes. Given the unique characteristics of each case, it is crucial to analyse specific legal documents, management structures and broader socio-political contexts to determine which approaches may be most effective.

The riverine rights cases discussed in this book demonstrate significant variation in their outcomes and the challenges they face. In the concluding chapter (Borchgrevink and Brandshaug this volume), a summary is made of the insights gained from these case studies, evaluating the extent to which they fulfil the promises and critiques of rights-based environmental protection. Additionally, we consider how these cases are positioned within the broader dynamics of global and local legal transformations, offering perspectives on the potential future trajectories of the rights of nature movement.

Notes

1 https://ecojurisprudence.org/initiatives/huddle-v-provincial-government-of-loja/, https://www.earthlaws.org.au/wp-content/uploads/2016/07/RON_Vilcabamba-Ecuador-Case-complete.pdf

2 https://ecojurisprudence.org/dashboard/?map-style=political
3 https://ecojurisprudence.org/dashboard/?map-style=political
4 https://www.earthlawcenter.org/our-mission
5 https://www.garn.org/about-garn/
6 https://www.rightsofnaturetribunal.org/about-us/
7 https://iucn.org/our-union/commissions/group/iucn-wcel-rights-nature-task-force
8 https://www.internationalrivers.org/about/mission-vision/
9 https://www.rightsofrivers.org/
10 https://www.garneurope.org/rights-for-aquatic-ecosystems-in-europe/
11 https://www.embassyofthenorthsea.com/
12 https://water-bodies.eu/manifesto/
13 https://foreignpolicy.com/2021/11/24/nature-person-rights-environment-climate-philosophy-law/?fbclid=IwY2xjawGW49RleHRuA2FlbQIxMQABHZYef9hcn4BsU5EI1dUPyfn6yyASOxfj-QAzXX60wEcnd5IYA_-BPUePbQ_aem_QSV_k8rxFgBDm6YDxlBS5Q
14 https://www.undp.org/sites/g/files/zskgke326/files/2023-01/UNDP-UNEP-UNHCHR-What-is-the-Right-to-a-Healthy-Environment.pdf
15 https://www.ohchr.org/sites/default/files/documents/issues/environment/srenvironment/activities/2024-04-22-stm-earth-day-sr-env.pdf
16 https://docs.un.org/en/A/RES/71/266
17 https://docs.un.org/en/A/72/175
18 https://docs.un.org/en/A/RES/79/210
19 https://www.eesc.europa.eu/sites/default/files/files/qe-03-20-586-en-n.pdf
20 https://www.europarl.europa.eu/RegData/etudes/STUD/2021/689328/IPOL_STU(2021)689328_EN.pdf
21 https://www.stopecocide.earth/press-releases-summary/european-parliament-urges-support-for-making-ecocide-an-international-crime
22 https://wedocs.unep.org/bitstream/handle/20.500.11822/43008/global_climate_litigation_report_2023.pdf?sequence=3
23 https://www.lse.ac.uk/granthaminstitute/wp-content/uploads/2024/06/Global-trends-in-climate-change-litigation-2024-snapshot.pdf

References

Agarwal, Bina. 1992. "The gender and environment debate: Lessons from India." *Feminist Studies* 18 (1): 119–158. doi:10.2307/3178217.

Anderson, Mark. 2024. *The rights of nature and the testimony of things: Literature and environmental ethics from Latin America*. Vanderbilt University Press.

Barad, Karen. 2003. "Posthumanist performativity: Toward an understanding of how matter comes to matter." *Signs: Journal of Women in Culture and Society* 28 (3): 801–831.

Bennett, Jane. 2010. *Vibrant matter: A political ecology of things*. Duke University Press.

Berry, Thomas. 1988. *The dream of the earth*. Sierra Club Books.

Berry, Thomas. 1999. *The great work: Our way into the future*. Bell Tower.

Berry, Thomas. 2001. "The origin, differentiation and role of rights." Paper presented at the Earth Jurisprudence Conference, 21–24 April at the Airlie Centre, Warrenton, Virginia. Accessed 4 March 2025. https://www.ties-edu.org/wp-content/uploads/2018/09/Thomas-Berry-rights.pdf.

Blaser, Mario. 2013. "Notes towards a political ontology of 'environmental' conflicts." In *Contested ecologies: Dialogues in the South on nature and knowledge*, edited by Lesley Green, 13–27. HSRC Press.

Boelens, Rutgerd, Arturo Escobar, Karen Bakker, Lena Hommes, Erik Swyngedouw et al. 2022. "Riverhood: Political ecologies of socionature commoning and translocal struggles for water justice." *The Journal of Peasant Studies* 50 (3): 1125–1156. doi:10.1080/03066150.2022.2120810.

Boelens, Rutgerd, Lena Hommes and Jaime Hoogesteger. 2024. "Ontological politics in river defence debates: Unpacking fields of contention in eco-centric and non-human turns." *Water Alternatives* 17 (3): 649–668.

Boyd, David R. 2011. *The environmental rights revolution: A global study of constitutions, human rights, and the environment*. UBC Press.

Boyd, David R. 2017. *The rights of nature: A legal revolution that could save the world*. ECW Press.

Bullard, Robert. 1993. "The threat of environmental racism." *Natural Resources and Environment* 7 (3): 23–26, 55–56.

Chakrabarty, Dipesh. 2019. "The climate of history in a planetary age." *Critical Inquiry* 35 (4): 197–222.

Constitución de la República de Ecuador. 2008. Registro Oficial 449. Ecuador.

Corte Constitucional de Colombia. 2016. Sentencia T-622 (Río Atrato Sujeto de Derechos).

Cosgrove, Denis. 1994. "Contested global visions: One-world, whole-earth, and the Apollo space photographs." *Annals of the Association of American Geographers* 84 (2): 270–294.

Crance, Jessica L., Alexander E. Bowles and Ann E. Garver. 2019. "Evidence for vocal learning in Killer Whales (Orcinus orca)." *The Journal of the Acoustical Society of America* 146 (6): 4296–4310.

Cullinan, Cormac. 2002. *Wild law: A manifesto for earth justice*. Green Books.

De la Cadena, Marisol. 2010. "Indigenous cosmopolitics in the Andes: Conceptual reflections beyond 'politics'." *Cultural Anthropology* 25 (2): 334–370.

De la Cadena, Marisol, and Mario Blaser, eds. 2018. *A world of many worlds*. Duke University Press.

Deleuze, Gilles, and Félix Guattari. 1987. *A thousand plateaus: Capitalism and schizophrenia*. Translated by Brian Massumi. University of Minnesota Press.

Department of Conservation (New Zealand Government). 2014. Te Urewa Act. *Public Act No* 51.

Descola, Philippe. 2013. *Beyond nature and culture*. Translated by Janet Lloyd. University of Chicago Press.

Douzinas, Costas. 2000. *The end of human rights: Critical legal thought at the turn of the century*. Hart Publishing.

Dunlap, Alexander, and Sian Sullivan. 2020. "A faultline in neoliberal environmental governance scholarship? Or, why accumulation-by-alienation matters." *Environment and Planning E: Nature and Space* 3 (2): 552–579. doi:10.1177/2514848619874691.

Dunnett, Oliver, Matthew Heffernan and Chris Thorpe. 2019. "'Earthrise' and the emergence of global environmentalism." *Environmental History* 24 (1): 131–156.

Ecojurisprudence.org. 2002. *Navajo Nation Code Annotated*. https://ecojurisprudence.org/wp-content/uploads/2022/02/US_Navajo-Nation-Code_278.pdf.

Escobar, Arturo. 2020. *Pluriversal politics: The real and the possible*. Duke University Press.

García Ruales, Jenny, Katarina Hovden, Helen Kopina, Colin D. Robertson and Hendrik Schoukens, eds. 2024. *Rights of nature in Europe: Encounters and visions*. Routledge.

Gilbert, Scott F., Jan Sapp and Alfred I. Tauber. 2012. "A symbiotic view of life: We have never been individuals." *The Quarterly Review of Biology* 87 (4): 325–341.

Haraway, Donna J. 2008. *When species meet*. University of Minnesota Press.

Haraway, Donna J. 2016. *Staying with the trouble: Making kin in the Chthulucene*. Duke University Press.

Hommes, Lena, Jeroen Vos and Rutgerd Boelens. 2023. "The need to acknowledge, study and engage with new water justice movements." *PLOS Water* 2 (5): e0000128. doi:10.1371/journal.pwat.0000128.

Ingold, Tim. 2000. *The perception of the environment: Essays on livelihood, dwelling and skill*. Routledge.

Kauffman, Craig M., and Pamela L. Martin. 2021. *The politics of rights of nature: Strategies for building a more sustainable future*. MIT Press.

Kohn, Eduardo. 2013. *How forests think: Toward an anthropology beyond the human*. University of California Press.

La Asamblea Legislativa Plurinacional. 2010. Ley de Derechos de la Madre Tierra. Ley 071 del Estado Plurinacional (Bolivia).

Latour, Bruno. 1993. *We have never been modern*. Translated by Catherine Porter. Harvard University Press.

Latour, Bruno. 2004. "Whose cosmos, which cosmopolitics?" *Common Knowledge* 10 (3): 450–462.

Law, John. 2004. *After method: Mess in social science research*. Routledge.

Law, John. 2015. "What's wrong with a one-world world?" *Distinktion: Journal of Social Theory* 16 (1): 126–139.

Lenton, Timothy M., et al. 2019. "Climate tipping points—Too risky to bet against." *Nature* 575 (7784): 592–595.

Lovelock, James E. 1979. *Gaia: A new look at life on earth*. Oxford University Press.

Maldonado-Torres, Nelson. 2007. "On the coloniality of being: Contributions to the development of a concept." *Cultural Studies* 21 (2–3): 240–270.

Malm, Andreas, and Alf Hornborg. 2014. "The geology of mankind? A critique of the Anthropocene narrative." *The Anthropocene Review* 1 (1): 62–69. doi:10.1177/2053019613516291.

Mathews, Andrew S. 2017. "Ghostly forms and forest histories." In *Arts of living on a damaged planet*, edited by Anna L. Tsing, Heather Swanson, Elaine Gan and Nils Bubandt. University of Minnesota Press.

May, James R. 2006. "Constituting fundamental environmental rights worldwide." *Pace Environmental Law Review* 23 (2): 401–480.

Messeri, Lisa. 2015. *Placing outer space: An earthly ethnography of other worlds*. Duke University Press.

Mignolo, Walter D. 2007. "Introduction: Coloniality of power and de-colonial thinking." *Cultural Studies* 21 (2–3): 155–167. doi:10.1080/09502380601162498.

Mohd. Salim v. State of Uttarakhand Case. 2017. The High Court of Uttarakhand.

Mol, Annemarie. 2002. *The body multiple: Ontology in medical practice*. Duke University Press.

Moore, Jason W. 2016. *Anthropocene or capitalocene? Nature, history, and the crisis of capitalism*. MP Press.

Næss, Arne. 1973. "The shallow and the deep, long-range ecology movement: A summary." *Inquiry* 16: 95–100.

O'Gorman, Emily. 2017. "Imagined ecologies: A more than human history of Malaria in the Murrumbidgee Irrigation Area, New South Wales, Australia, 1919–45." *Environmental History* 22 (3): 486–514. https://www.jstor.org/stable/26155429?seq=1.

Plumwood, Val. 1993. *Feminism and the mastery of nature.* Taylor and Francis. https://www.taylorfrancis.com/books/mono/10.4324/9780203006757/feminism-mastery-nature-val-plumwood.

Quijano, Aníbal. 2000. "Coloniality of power, eurocentrism, and Latin America." *Nepantla: Views from South* 1 (3): 533–580.

Rockström, Johan, et al. 2009. "A safe operating space for humanity." *Nature* 461 (7263): 472–475.

Shapiro, Martin, and Alec Stone Sweet. 2002. *On law, politics and judicialization.* Oxford University Press.

Sheldrake, Merlin. 2021. *Entangled life: How fungi make our worlds, change our minds, and shape our futures.* Random House.

Shelton, Dinah. 2020. *Advanced introduction to international environmental law.* Edward Elgar Publishing.

Shiva, Vandana. 1989. *Staying alive: Women, ecology and development.* Zed Books.

Simard, Suzanne W., Kevin J. Beiler, Marcus A. Bingham, Julie R. Deslippe, Leanne J. Philip, François P. Teste. 2015. "Mycorrhizal networks: Mechanisms, ecology, and modelling." *Fungal Biology Reviews* 29 (2): 39–57.

Smith, Linda T. 1999. *Decolonizing methodologies: Research and indigenous peoples.* Zed Books.

Steffen, Will, Paul J. Crutzen and John R. McNeill. 2007. "The Anthropocene: Are humans now overwhelming the great forces of nature?" *The Anthropocene BIO: A Journal of the Human Environment* 36 (8): 614–621.

Steffen, Will, Wendy Broadgate, Lisa Deutsch, Owen Gaffney, Cornelia Ludwig. 2015. "The trajectory of the Anthropocene: The great acceleration." *The Anthropocene Review* 2 (1): 81–98.

Stengers, Isabelle. 2010 [1997]. *Cosmopolitics I.* Translated by Robert Bononno. University of Minnesota Press.

Stone, Christopher D. 1972. "Should trees have standing? Towards legal rights for natural objects." *Southern California Law Review*: 450–493.

Swanson, Heather, Anna L. Tsing, Nils Bubandt and Elaine Gan. 2017. "Introduction: Bodies tumbled into bodies." In *Arts of living on a damaged planet*, edited by Anna L. Tsing, Heather Swanson, Elaine Gan and Nils Bubandt. University of Minnesota Press.

Tanasescu, Mihnea. 2013. "The rights of nature in Ecuador: The making of an idea." *International Journal of Environmental Studies* 70 (6): 846–861.

Tanasescu, Mihnea. 2022. *Understanding the rights of nature: A critical introduction.* Transcript Verlag.

Tanasescu, Mihnea, Elizabeth Macpherson, Davis Jefferson and Julia Torres Ventura. 2024. "Rights of nature and rivers in Ecuador's Constitutional Court." *The International Journal of Human Rights.* doi:10.1080/13642987.2024.2314536.

Taylor, Andrea H., and Matthew L. Cody. 2019. "Avian cognition: Exploring the intelligence of birds." *Current Biology* 29 (10): R390–R392.

Thiong'o, Ngugi wa. 1986. *Decolonising the mind: The politics of language in African literature.* Heinemann.

Todd, Zoe. 2016. "An indigenous feminist's take on the ontological turn: 'ontology' is just another word for colonialism." *Journal of Historical Sociology* 29 (1): 4–22.

Tola, Miriam. 2016. "Composing with Gaia: Isabelle Stenger and the feminist politics of the earth." *PhaenEx* 11 (1): 1–21. doi:10.22329/p.v11i1.4390.

Tsing, Anna L. 2013. "More-than-human sociality: A call for critical description." In *Anthropology and nature*, edited by Kirsten Hastrup, 27–42. Routledge.

Tsing, Anna. 2015. *The mushroom at the end of the world: On the possibility of life in capitalist ruins.* Princeton University Press.

Viveiros de Castro, Eduardo. 2004. "Perspectival anthropology and the method of controlled equivocation." *Journal of the Society for the Anthropology of Lowland South America* 2 (1): 3–20.

Yazzy, Melanie K., and Cutcha Risling Baldy. 2018. "Introduction: Indigenous peoples and the politics of water." *Decolonization: Indigeneity, Education & Society* 7 (1): 1–18.

9 Beyond riverine rights

Malene K. Brandshaug and Axel Borchgrevink

Introduction

In the first chapter of this book – "Exploring riverine rights" – we posed three sets of questions underlying the research project on which this book is based:

- Why did these legal innovations for rivers happen, and why at virtually the same time in three countries as different as Aotearoa New Zealand, Colombia and India?
- What have been the outcomes of granting legal personhood to rivers?
- What can we learn from these cases?

In the book, we have examined each case in detail – Cribb on the Whanganui River in Aotearoa New Zealand Vallejo Piedrahíta, McNeish, Moreno Socha, Cuesta Mosquera and Alfonso León on the Atrato River in Colombia; and Borchgrevink and Upadhyay on the Ganga and Yamuna rivers in India. Moreover, we have analysed the relations between people and their rivers in each of these cases (Brandshaug, Cribb, Moreno Socha, Alfonso León and Cuesta Mosquera); the justifications for and implications of the legal innovations (Vallejo Piedrahíta, Montoya-Restrepo, Arango Duque and Gaviria Mira); the politics involved in these legal decisions and their implementation (Borchgrevink); and how these cases reflect international campaigns for rights of nature and broader global trends (Brandshaug and McNeish).

In this chapter, we return to the three questions and use the data and analyses presented to address them. We argue that the way forward must be about looking *beyond* legal personhood and rights for rivers.

Why? Why now?

This book has demonstrated that riverine rights is not a uniform phenomenon; the cases of granting personhood rights to Whanganui, Atrato, Ganga and Yamuna are very different. Here, we therefore draw on the particularities of the individual cases in explaining *why* these legal innovations for rivers happened. However, these riverine rights cases are not only locally embedded;

DOI: 10.4324/9781003581208-9

Figure 9.1. Headwaters of the Whanganui River. Photo by Axel Borchgrevink.

we also want to address important similarities they have to each other and to global movements and dynamics.

Whanganui

Whanganui River is of pronounced significance for Māori of Aotearoa New Zealand, particularly the Whanganui *iwi* (tribe). Historical and current experiences of land and river deprivation caused by colonialism and government policies have led to persistent efforts from Whanganui *iwi* to safeguard their rights and sovereignty over the river and to ensure the river's health and vitality. The Treaty of Waitangi, signed in 1840, granted rights to the Māori but was rarely adhered to in practice. One result of colonial appropriation of territory – which in most instances was motivated by economic interests – was environmental degradation and exploitation of the river. From the first river claim by the Whanganui Māori to Parliament in 1873, more than a hundred years of legal battles and negotiations between Māori groups and the New Zealand government followed. The rise of the Māori rights movement led to the Waitangi Tribunal being established in 1975 to investigate and make recommendations on claims brought by Māori individuals, groups or organizations regarding land and water confiscation, breaches of the Treaty of Waitangi and other injustices towards Māori. Over the years, the Waitangi Tribunal has facilitated numerous settlements to rectify historical grievances and uphold Indigenous rights (Cribb this volume).

The Te Awa Tupua (Whanganui River Claims Settlement) Act that granted the Whanganui River legal personhood in 2017 was a significant milestone in this ongoing process. The Act acknowledges the river as a living entity with rights and includes Māori principles and values regarding their inseparable relationship to the river – emphasizing the importance of acknowledging and protecting Indigenous rights within Aotearoa New Zealand's legal framework. As such, it can be seen as part of efforts to create a more inclusive state by devolving authority to the Whanganui Māori (Borchgrevink this volume). The *iwi*'s deep relationship to the river and their concern over its deteriorating ecological state along with repeated historical injustices towards the Māori of Aotearoa New Zealand have together driven the process of securing the protection of the river and Whanganui *iwi* rights to have a say in its management.

Hence, the process leading up to the passing of the Te Awa Tupua Act was intertwined with the long history of Māori political activism and its larger historical, political, cultural and socio-legal context. The idea of granting personhood rights to the river, as an alternative to the Māori claims for ownership of the river, emerged in the negotiations over a settlement of the Whanganui claim. The government negotiator John Wood had knowledge of and was interested in legal models that recognize personhood for natural entities, and Chris Finlayson, the Minster for Treaty Settlements, was intrigued by the idea of legal personhood and legal precedents elsewhere. The *iwi*'s proposal was for a settlement that emphasized their special relationship to the river. The parties jointly agreed that personhood was the best legal vehicle for this recognition (Cribb this volume, Finlayson and Christmas 2021). The passing of the Te Awa Tupua Act aimed to make up for historical grievances and promote reconciliation by including Whanganui *iwi* concepts of human-environment relationships and caring river management. In addition to the enduring mobilization and activist efforts of the Whanganui *iwi*, the legal decision happened in the context of a larger effort to address past injustices and of growing global attention to environmental issues and Indigenous rights recognition.

Atrato

In 2016, the Atrato River was granted the status of legal subject in response to severe environmental degradation and pollution. The legal decision was driven by concerns over the river's poor ecological condition and the objective of recognizing its importance to local communities. The Atrato River has long suffered from severe environmental degradation caused by the effects of gold mining, mercury pollution, deforestation and other extractive activities, which has resulted in a significant decline in water quality and biodiversity. These issues have directly impacted Indigenous and Afro-Colombian communities particularly, who depend on the river for their livelihoods and cultural practices. Contaminated water not only significantly harms the river's health but has also led to increased health problems in the human population and poses a threat to their way of life (Vallejo Piedrahíta, McNeish, Moreno Socha, Cuesta Mosquera and Alfonso León this volume). Atrato is located in a region with high levels of violence, illegal economic activities and limited state presence, which has allowed the continuation of the large-scale destruction of ecosystems and the resulting contamination and health problems (Borchgrevink this volume).

Environmental organizations and activists from local communities have been crucial in raising awareness about the river's plight and advocating for stronger legal protections to ensure its restoration and conservation. In 2015, the Centre for Social Justice Studies (Tierra Digna) filed a lawsuit on behalf of local communities, arguing that the serious environmental issues of the Atrato River were violating the communities' constitutional rights to life, health, water, food security and a healthy environment as well as their cultural and territorial rights. Colombia's Constitutional Court recognized the gravity of the situation and reacted by handing down the 2016 Sentencia T-622, which acknowledged the violation of the above-mentioned rights and additionally introduced the category of "biocultural rights" and recognized the Atrato River as a legal subject with rights to protection, conservation, maintenance and restoration. Hence, the ruling not only aimed to restore the river's ecological balance but also safeguard the cultural and economic well-being of the communities dependent on it. By granting them a significant role in the river's management and conservation efforts, it also sought to empower these communities. Moreover, the decision to go beyond proclaiming a systematic violation of human rights and to declare Atrato River as a subject with rights was inspired by the legal innovation of granting personhood rights to Whanganui River. National and international legal precedents also laid the ground for the ruling, such as ILO Convention 169 (1989) on the rights of Indigenous and Tribal Peoples, and Colombia's 1991 "ecological" constitution and Law 70 of 1993 recognizing the ethnic and territorial rights of Afro-Colombian communities (Vallejo Piedrahíta, McNeish, Moreno Socha, Cuesta Mosquera and Alfonso León this volume).

The legal process was initiated by local communities and organizations, while the court in seeking innovative ways to address socio-environmental crises responded with radical measures that went beyond their petition, in order to promote sustainable and equitable management of the river. Global awareness of the urgency of environmental issues likely also played a role in the timing of the decision, moving state actors to take more decisive action.

Ganga and Yamuna

The Ganga and Yamuna have for a long time been subject to increased encroachments and disruptions of their water flows caused by large-scale damming for hydroelectric production, canals, irrigation, infrastructure development, sandmining and extraction of groundwater. Coupled with pollution from industrial, agricultural and domestic activities, and with the impacts of climate change, this has caused severe environmental problems. The degraded state of the river and its banks have led to campaigns and protests by environmental activists, religious organizations and riverine populations for decades. Despite long-term awareness of these issues, political and legal action has so far proven inadequate in sufficiently addressing them.

Two public interest litigations were filed in 2014 and 2015, seeking injunctions against river encroachments, mining and contamination (Borchgrevink this volume). In response, the judge Rajiv Sharma of Uttarakhand High Court granted Ganga and Yamuna the status of legal persons in March 2017. While the ruling was influenced by environmental concerns and public outcry over the state of the rivers, the radical decision of giving legal personhood and rights to the rivers came as a surprise. It had not been asked for in the original petition and did not follow logically from the issue to be addressed. Moreover, there had been little public interest in the concepts of legal personhood and rights for nature in India. A significant factor behind the legal decision was the judge, who has been known for his proactive stance on environmental issues and willingness to make unconventional rulings. Justice Sharma's commitment to environmental conservation and knowledge of other legal models for granting personhood rights to natural entities – such as the process of recognizing Whanganui River as a legal person – is likely to have played a crucial role in the court's decision to grant legal personhood to the rivers. The ruling also reflects a broader trend of judicial activism in India, a liberal attitude to public interest litigations and the growing global shift towards recognizing the rights of nature (Borchgrevink and Upadhyay this volume).

Although the ruling is primarily aimed at improving the ecological state of the river and combating environmental threats towards it, the decision is also linked to the significance of the Ganga and Yamuna for Hindus and for India. In the ruling, the concept of legal personhood is coupled with ideas of the rivers as living entities. And the spiritual significance of the rivers for Hindus, as well as the close connection between people and rivers, is emphasized.

In-between global flows and local particularities

Granting rights to nature and to rivers is often seen as a global phenomenon. The riverine rights cases discussed in this book appeared at almost the same time, which can give the impression of there being a strong link between them. However, we have not found any coordinated efforts connecting the cases, and the main drivers behind the legal decisions are local. The links between the cases are rather weak, although links do exist and are important. The cases are not only a result of local particularities, isolated from connections to other cases, ideas and larger global dynamics; they are located somewhere in-between global currents and local processes (Kinkaid 2019).

In all cases – in India, in Colombia and in Aotearoa New Zealand – there were local movements for change behind the legal riverine rights decisions.

Local people living with the rivers in question have mobilized – and in some cases formed alliances with other actors – to address the gravity of extractive activities and environmental issues related to the river and/or injustices towards its riverine people. That is not to say that the particularities of the legal decisions themselves are a direct result of a grassroots movement. In India, for instance, the relationship between local claims and wishes for the rivers and the content of the legal decision is weak. The decision to grant personhood rights to Ganga and Yamuna is clearly top-down. In the case of Colombia, the broadly supported local initiative to address serious socio-environmental problems through stronger legal protection is more directly a factor that contributed to the T-622 ruling. However, in their petition, the local communities only asked for protection of their human rights, rather than for legal personhood for the river. In the case of Aotearoa New Zealand, however, the local grounding of the Te Awa Tupua Act is quite unique, both regarding the significance of long-term Māori activism and legal battles through the Waitangi Tribunal but also Whanganui *iwi* influence in developing and shaping the Te Awa Tupua Act, including incorporating Māori principles and values towards the river, and their further role in its implementation. Moreover, in this case, the riverine rights were established through legislation, in contrast to the judicial path in the two other cases. Whether top-down or grassroots, the legal decisions in all cases were initiated by actors within each country, rather than having been pushed by international organizations or movements.

While the social origins and the problems to be addressed in the three cases may share some similarities, they are quite different due to the different contexts in which they emerge, in terms of, for example, politics, culture, economy, ethnic composition, institutional structures and available legal mechanisms. Interestingly, though, they all emerge within (post)colonial states, are concerned with questions of sovereignty in different ways, are shaped by a political economy of extractivism and address ethnic and cultural complexities (Vallejo Piedrahíta, Montoya-Restrepo, Arango Duque and Gaviria Mira this volume, Borchgrevink this volume). Moreover, they are united by being cases of riverine rights occurring at the same time, thereby forming part of a growing rights of nature movement. The concept of granting legal personhood to rivers has been developing across countries, beginning with these three cases in Colombia, Aotearoa New Zealand and India. The ideas behind riverine rights travel; they are inspired by and inspire similar developments elsewhere. In these processes, they are transformed and adjusted as they are transplanted into new contexts. In the three cases discussed here – although very different and primarily locally driven – there are examples of cross-fertilization and links between cases and to legal debates and precedents elsewhere.

In Aotearoa New Zealand, as mentioned above, the government negotiators were inspired by Stone's seminal paper "Should trees have standing" (1972) and were interested in legal personhood models for natural entities.

Legal examples that influenced the decision include the Waikato River being represented by a guardian and the idea of legal personality of *taonga* (prized possession), as well as the legal status of temples and idols in India (Cribb this volume, Finlayson and Christmas 2021).

The Whanganui River case in Aotearoa New Zealand in turn inspired Colombia's Constitutional Court's recognition of the Atrato River as a subject of rights, thus going beyond a focus on human rights violations and emphasizing the interrelatedness of rivers and riverine communities (Vallejo Piedrahíta, McNeish, Moreno Socha, Cuesta Mosquera and Alfonso León this volume). In what Macpherson and Clavijo term "an outward looking approach to Colombian constitutional development", the Court explicitly references the Whanganui and Te Urewera cases, as well as the Ecuadorean 2008 Constitution, which included rights of nature or Pachamama and Bolivia's recognition of the rights of Mother Earth in 2010 (Macpherson and Clavijo Ospina 2018, 290).

Also the Uttarakhand High Court's decision to grant legal personhood to the rivers Ganga and Yamuna was influenced by the Whanganui case, which was brought to the attention of the court by the lawyer of one of the petitioners. The ruling mirrored phrases from the Te Awa Tupua Act, showing a direct influence. The glacier ruling granting personhood status to the sources of the Ganga and Yamuna that followed shortly after the river ruling is even more explicit about its international sources of inspiration. It refers to articles on the social and spiritual value of nature, natural science texts on biodiversity, international conventions and declarations, passages from the Chipko movement, the Tilari declaration, the Te Urewera case in New Zealand, discussions of the *parens patriae* principle in the US and the constitutional legal rights of "Mother Earth" (Borchgrevink and Upadhyay this volume).

Apart from these concrete links and explicit inspirations, the development of riverine rights at multiple places at more or less the same time – and the formation of a larger rights of nature movement – can be understood in relation to a number of global dynamics. A growing recognition of environmental degradation and the urgent need for sustainable practices have been driven by a heightened awareness of climate change, biodiversity loss and the harmful impacts of industrial activities on ecosystems. Here, scientific research has played a crucial role in uncovering the causes of the multiple environmental crises of our time and in emphasizing the complex interconnections between humans and the environment. These interrelationships have also been a central theme in social theory – such as posthumanism, feminist materialism and decolonial thought – forming critiques of modernist practices, power hierarchies and Enlightenment thought based on strict separations between nature-culture, body-mind, thing-being, etc. Furthermore, influenced by larger movements addressing environmental and Indigenous injustices, international agreements and the work of international non-governmental organizations have further propelled the discourse on environmental rights, emphasizing the interconnectedness of human wellbeing and the health of ecosystems. In this larger move, rights of nature and rights of rivers activism have recently been especially pronounced, advocating for a paradigm shift towards viewing nature as a rights-bearing entity deserving of legal protection and stewardship. This larger movement is often explicitly inspired by Indigenous environmental ethics and practice (Brandshaug and McNeish this volume).

These are some of the characteristics of the particular moment in time where three countries almost simultaneously recognized the personhood rights of rivers. A special conjuncture, involving a combination of local and global developments and events, meant that time was ripe for such attempts at ecocentric lawmaking. Also, in all three cases, rivers seemed to be a good vehicle for the granting of rights: rivers combine many realms of human life (cultural, material, practical, social) and also the life of the larger ecosystem with its many species (Brandshaug et al. this volume).

Outcomes

Whanganui

The immediate and most evident outcome of the Te Awa Tupua (Whanganui River Claims Settlement) Act was that the more than 140 years of formal claims and litigation by the Whanganui *iwi* for sovereignty over the river was recognized and accepted as legitimate by the Crown. The Act includes a section where the Crown apologizes for having "undermined the ability of Whanganui Iwi to exercise their customary rights and responsibilities in respect of the Whanganui River, and consequently the expression of their mana".[1] By having the *Tupua te Kawa* – Māori law or values – as its fundamental principles, the Act is a strong example of legal pluralism, where Indigenous law is given prominence within formal state law.

Thus, a direct outcome, deriving from Parliament approving the law, was an explicit recognition of the Indigenous rights of the Whanganui Māori. Furthermore, the inclusion of Māori law – the *Tupua te Kawa* – into the Act, and the insistence that *Te Awa Tupua* is "a living and indivisible whole, comprising the Whanganui River from the Mountain to the sea, [and] incorporating all its physical and metaphysical elements" is a new and incremental step in the ongoing "legal revolution" (Cribb et al. 2024, 6) of Aotearoa New Zealand, where Māori law is given increasing recognition within the legal system. Recognition of Indigenous rights and the deepening of legal pluralism in Aotearoa New Zealand are indisputable among the outcomes of this new piece of legislation.

In our study, we have also explored the changes the Act has led to in governance and environmental management (Cribb this volume, Cribb et al. 2024). The law has established new institutions and procedures. In two cases we have studied – the committee tasked with developing a new strategy for the health and wellbeing of the river basin, and the project for restoring the port of Whanganui – we find that substantial efforts are made to involve those affected by these programmes, and to devolve decision-making to *iwi, hapū* (tribes and sub-tribes) and local communities of Māori and non-Māori. The Act thereby promotes participatory democratic procedures.

Furthermore, these new institutions are structured to ensure that decisions are made considering the river as a whole, and with a view to promoting the river's health and "restoring its abundance". Of course, decision-making often involves difficult deliberations involving the balancing of competing interests, and there are no guarantees that nature's interests will be prioritized in all future cases. But it seems evident that the new institutions, with their holistic approach, offer a better chance of decisions being made that are good for the environment and for the river than the system it has replaced, which tended to break down the river into parts – water, banks, riverbed, transport, fishing, irrigation, bridges and roads, etc. – that were dealt with separately (Cribb, this volume).

Yet, as this book has also documented, there are limits to what the Te Awa Tupua Act can do. The Act specifically states that existing property rights or rights to water, fish, animal or plants are not affected by the law.[2] Thus, most of the activities that have led to the deterioration of the river are allowed to continue. Even if the Act has sparked new initiatives for restoring the health and wellbeing of the river,[3] the short-term environmental effects of the Act are limited.

Atrato

The ruling T-622/16 established a new concept in Colombian jurisprudence: biocultural rights, to reflect the close interconnections and interdependence of the Atrato River and the Indigenous and Afro-Colombian communities living along it. And by declaring the river a legal subject with the rights to protection, conservation, maintenance and restoration, the Constitutional Court made a significant step in Colombia's move towards recognizing rights of nature (Vallejo Piedrahíta, McNeish, Moreno Socha, Cuesta Mosquera and Alfonso León this volume). These legal innovations were direct outcomes of the ruling. They have inspired 17 subsequent cases of rights for nature granted in Colombia (Vallejo Piedrahíta, McNeish, Moreno Socha, Cuesta Mosquera and Alfonso León this volume, Table 3.1).

In the region of Chocó, and in the communities along the river, there is satisfaction over the ruling having recognized the serious breaches of human rights that the contamination of the river, the deforestation and the state absence result in. People also express that the ruling gives them a new platform for making claims and interacting with the institutions of the state. The guardians appointed to represent the riverine communities is a new institution with a formal position to be heard. And the granting of legal subjecthood and biocultural rights provides new ways of framing normative arguments for the urgency of addressing the environmental situation, and for protecting their way of life.

Yet, there are limitations to the ruling in that it only addresses illegal mining and not large-scale licensed mining and logging, which also cause environmental damage. And, in terms of implementation of the orders to clean up the river and stop illegal gold mining, there are significant challenges. They include weak institutional coordination, insufficient funding, conflicting understandings and objectives and, at times, insufficient political backing. In sum, these weaknesses mean that there has been no reduction in contamination of the river, and that illegal mining, lawlessness and violence continue unabated (McNeish and Moreno Socha 2024).

The Ganga and Yamuna

The two rulings that granted legal personhood to the Ganga and Yamuna Rivers (Mohd Salim v. State of Uttarakhand 2017) and to Uttarakhand glaciers, rivers, jungles, meadows and wetlands (Lalit Miglani v. State of Uttarakhand 2017) were stayed after a few months. They remain suspended until the Supreme Court of India processes the appeals against the rulings. Implementation of the legal personhood decisions was therefore halted before it had really started. The effects the rulings have had in terms of changes in environmental governance and management are consequently negligible. The state of the rivers cannot be said to have been improved by these decisions. Moreover, the judgments did not follow from social movement initiatives and did not have any significant social backing or inspire any great popular enthusiasm. The impact of these rulings on popular imagination is thus also limited.

In sum, then, the effects of these rulings for the Ganga and Yamuna rivers and the people living along them have been minimal but have nevertheless rippled through the Indian judiciary, with subsequent cases that have accorded legal personhood and rights to natural entities. Some of them have been written by the same judge, a few by others. Moreover, in 2017, the legislative assembly of Madhya Pradesh passed a resolution granting legal personhood to the Narmada River, reportedly after inspiration from the Ganga and Yamuna ruling (Borchgrevink and Upadhyay this volume). And possibly the Supreme Court will at some point deliberate on the cases, which could be a new opportunity for rights of nature to gain traction within India's legal system.

Outcomes for rivers – eight years on

A general conclusion on outcomes is that they can be very long in the making. When we designed this project, we thought that starting our study in 2020, three years after the legal decisions of legal personhood for rivers, would be a good time for us to see and understand the impacts they would lead to. This was quite naive. At the time of writing, eight years on, it is still difficult to determine what the effects of these legal innovations will be in these three countries. In India, the Supreme Court has not opened the process of reviewing the rulings, and there are no signs that it will do so in the near future. Until that happens, we won't know whether there will be the potential for any direct practical effects from the implementation of the rulings. In Colombia, outcomes in terms of resolving the issues of contamination, extractivism and deforestation that the claimants raised in their petition to the Constitutional Court might still be achieved, even if so far they have not. And in Aotearoa New Zealand, achieving the objectives of restoring "the abundance" of the Whanganui River is something that will require decades if not centuries, and while signs so far are positive, practical steps to achieve these objectives are still only emerging.

This quick review of the impacts that these legal decisions have had repeats the finding that resonate throughout this book: these cases are widely dissimilar. The rivers, the problems to be addressed, the contexts and the scales are different. And while some form of legal personhood is common to the three cases, the legal mechanisms and institutions created for their implementation are different. It should therefore not be surprising that also outcomes vary.

The Whanganui case is clearly the most promising. Its primary achievements are in terms of responding to legitimate Māori claims, granting Indigenous rights and introducing more participatory forms of governance. There is a likelihood that over time the new institutions and mechanisms will also lead to better environmental outcomes. But since existing property rights and ongoing activities are not affected, there are limits to how transformatory the new law can be. The Colombia case is the in-between case, where positive impacts are found in terms of recognition of community representatives and the legitimacy of their claims, as well as in subsequent legal decisions on rights of nature. In practice, though, the T-622 ruling has not led to improvement in the lives of the communities living along the river, nor in cleaning the river or halting deforestation and lawlessness. While in India, due to the rulings having been stayed shortly after they were made, impacts are few.

This raises the question of why the recognition of legal personhood for rivers has such unequal effects. What explains the success of the Whanganui case when compared to the other cases? We must be wary of drawing hard conclusions about this from the small sample of three cases. Yet, it is possible to reflect on the mechanisms that lead to positive impacts in one case and hinder them in another. The next section presents our conclusions in this respect.

The future of riverine rights

When we were planning this research project, we were motivated by curiosity over why apparently the same thing had happened almost simultaneously in three countries as different as Aotearoa New Zealand, Colombia and India. However, as we started our study, it quickly became clear that the underlying assumption was questionable. In important ways, what had happened in these three countries was not "the same thing". Our socio-legal approach – studying these legal innovations in their total social contexts – made us particularly sensitive to the fact that what we were dealing with were three very dissimilar cases. It forced us to reflect on what our planned comparative analysis of them could achieve. If these cases were so different, what could be gained by comparing them?

This question points to one very clear conclusion from the study: *These experiences do not involve blueprints or models that can be extracted, transferred and replicated or applied in other contexts.* Each case of legal personhood is unique having emerged from locally specific objectives and been shaped by its particular context. As socio-legal studies have amply demonstrated, law is always embedded in its social context.

This is not to say that these cases can offer no lessons or insights that can be of use and interest for other contexts. By examining the social mechanisms and processes that explain the different experiences and outcomes, learnings of a different order from a blueprint model can be gained. Moreover, even if different, the cases can be compared. By juxtaposing them, we find that there are also some commonalities. Both differences and commonalities allow for further reflections on what can be learnt from these cases. The remainder of this chapter draws out such general conclusions from our material.

The commonalities of these cases should be briefly pointed out. These three countries are all complex societies made up of different ethnic groups. They all share a common history of having been colonized by European states. While the colonial experience varies between them (see Borchgrevink this volume), in each case European legal systems have been imposed upon pre-existing legal and normative orders. The three cases of legal personhood all include elements of seeking to bridge such different orders through legal pluralism. Furthermore, even if not necessarily the primary intention, the three legal innovations all relate to the global discourse on the rights of nature, and they have created some form of rights for nature. And, in each case, it is rivers that are given rights.

From our study, the Whanganui experience emerges as the most successful. It has introduced a participatory form of management, where authority is devolved to local levels and Māori and non-Māori are involved in discussing and deciding on how to manage initiatives that affect the river. Principles for management are explicitly based on *kawa* – Māori values or law – which emphasize the inseparability and mutual dependence of the river and its people. Decisions should be made with reference to a holistic understanding of river, people and the environment as a whole. The Te Awa Tupua Act details a set of new institutions to be established, and others have been created for its implementation. There is funding for these institutions, and local government is cooperating in the process (Cribb et al. 2024). In sum, the Te Awa Tupua Act has been successful by combining *broad popular participation, local attachment to the river, institutional innovation and getting public institutions on board.*

These are elements that would be useful to try to emulate if similar legal arrangements are to be attempted elsewhere. There is no recipe for achieving this successful combination. But it is useful to look at the drawn-out process leading up to the Act. The Whanganui *iwi* filed their claim in 1990. In 1999, the Waitangi Tribunal concluded that the Treaty of Waitangi had guaranteed Whanganui *iwi* the rights to ownership, authority and control of the river, an ownership which had never been willingly relinquished. Negotiations between Government and the Whanganui *iwi* took place between 2002 and 2004, and again from 2009 until the settlement was signed in 2014. These last years were particularly important, when dedicated and wise negotiators from both sides forged the details of the agreed settlement, which basically included all the elements to go into the Act that was approved three years later. The patience and willingness not to give up, the thoroughness and attention to detail and the creativity that went into this process were remarkable. This allowed the creation of the successful Te Awa Tupua model. Furthermore, since the Act was approved, similar patience and wisdom have been demonstrated. The way in which consensus has been sought between all stakeholders – Māori and non-Māori, civil society and public institutions – in *Te Kōpuka*, the body tasked with developing a management strategy for the river, has been time-consuming but has ensured the anchoring of the process among all sectors of society (Cribb et al. 2024, Cribb this volume).

There are challenges to making this model function. It takes time to arrive at a consensus and conclusions. Processes are drawn out, and projects become more costly. Still, people we spoke with pointed out that these costs were more than outweighed by the improved way of doing things and the consequent savings in the long run. But there is of course the risk that this attitude may not be shared by everybody all of the time. Moreover, this approach is not only time-consuming but also labour intensive. Even if there is funding for some positions, there is also a major input of voluntary work. It puts a significant strain specifically on leaders and activists of the Whanganui *iwi*, who have a great responsibility for guiding the process and ensuring the broad involvement not only of Māori but of all people living in the river basin.

When comparing the processes through which the legal personhood arrangement was arrived at in the three cases, the Whanganui settlement stands out. Through drawn-out negotiations, the nitty-gritty institutional details were painstakingly developed and agreed between the parties, local Indigenous law – the *kawa* – was integrated into a parliamentary Act and legal personhood was agreed as a compromise to resolve the difficult issue of ownership. This contrasts starkly with the process in India, where the radical innovations of legal personhood did not respond logically to the claims of the litigants but were the personal creation of a judge who was known for his speed in coming up with rulings that were often unconventional. The Colombia case falls somewhere in between, as it started with a community-driven petition to the Constitutional Court and the process thus had a clear social grounding. On the other hand, the innovations of legal personhood and biocultural rights were introduced in a top-down manner. If the Whanganui case appears more promising than the other two, it seems likely that this is in part due to the patient and inclusive process leading up to the settlement and the Act.

Another thing that jumps out when comparing the cases is how different the three states are in terms of sovereignty over their territory and capacity to implement their policies (Borchgrevink this volume). Aotearoa New Zealand is a well-functioning state with an effective bureaucracy, control over its territory and few challenges to the rule of law. India, in contrast, has a state apparatus that is widely perceived to be bureaucratic and ineffective, as well as prone to elite capture. Colombia has limited sovereignty over large swathes of the country including remote areas like the Chocó region, where the Atrato River flows. Illegal goldmining, organized crime and guerilla and paramilitary groups can operate in these areas with relative impunity. While both India and Colombia are known for having strong and independent judiciaries, the impact of decisions that these courts make is clearly limited by the lack of capacity and effective sovereignty of the state apparatus in these two countries.

It is probably an exaggeration to say that having a state as effective as Aotearoa New Zealand is a precondition for rights of nature to have any effects. But state capacity matters. One question that emerges from reflecting on the three cases can be put as follows: if states are not complying with existing legislation for environmental protection, why should we expect another outcome when we introduce a radically new legal principle – rights of nature – that is challenging to implement and may conflict with other legislation? This is a rhetorical question, not one that needs to be answered. But it points to the fact that state capacity, the rule of law and sovereignty are preconditions that will influence the potential for rights of nature to have a direct impact.

Furthermore, a related issue needs to be taken into account: the political will to implement measures for environmental protection when they are in conflict with significant economic interests. This also points to larger political-economic structures that may hamper the transformative force of riverine rights legislation: the global capitalist economy that relies on growth and resource extraction and tends to deepen social inequalities.

Summing up the general lessons so far, our findings indicate that the following factors favour the establishment of a riverine rights regime that works:

- Allowing for inclusive participation and time in the process of developing the details of institutionalizing rights of nature
- Ensuring a framework that reflects local knowledge and values and promotes broad participation
- Having a state with sufficient sovereignty, capacity and will to enforce its environmental protection policy

In short, law is socially and culturally embedded, and the legal must be seen in conjunction with the political. Our study shows clearly that the *legal* phenomenon "personhood rights for rivers" is *intrinsically political*. Despite the often-reported trend towards judicialization of politics – that what was formerly solved through political means is now increasingly solved through legal means – one can argue that politics still tend to trump the legal. But perhaps it is better to view the opposition as false and that in modern states the legal and the political, the formal and the informal, inevitably go together. A too narrow focus on the *legal* side of these innovations might end up overlooking the political dimensions of riverine rights.

It has been argued that rights of nature is really about distributing power among people (Tanasescu 2022) but also that there is an asymmetrical power relation between recognizer and recognized where rivers are recognized as subjects within state-dominated legal frameworks. This recognition has the potential to be empowering, used as a political tool for making claims, but could also be a form of subjectification in a Foucauldian sense – of defining and controlling (Boelens et al. 2024, 661). Moreover, it has the potential to be co-opted by powerful actors and to strengthen instead of breaking down current power hierarchies (Brandshaug and McNeish this volume). To be aware of the political sides of riverine rights can help uncover the power dynamics caught up in rights recognition. Tanasescu (2022) argues that the crucial issue in this regard is who gets to represent nature. In the cases examined in this book, different types of guardians or institutions designed to speak for the river are created. In the case of Whanganui and Atrato, local communities are strongly represented in the guardian bodies and institutions that somehow represent the river. The intention in both cases is to empower these groups. As funding and institutional coordination is better in Aotearoa New Zealand, this intention is realized to a greater extent here. In India, in contrast, the heads of state institutions were named as guardians, thus pointing to a different type of politics that one would not expect to strengthen the position of disempowered people.

As discussed above, one thing that can be learned from these three cases of granting personhood rights to rivers is that, more than the legal concept of personhood rights in itself and the granting of rights to rivers per se, perhaps the most transformative potential lies in creating new forms of *participatory* governance and lawmaking, where local people with close relationships to, lived experiences with and knowledge of the river are included and have real power in decision-making – facilitating legal pluralism and reflecting local lives and ideas (Cuesta Mosquera and Vallejo Piedrahíta 2024). This kind of lawmaking and river governance is difficult and complex work, involving an interweaving of different knowledges, worldviews, ontologies, values, norms and/or laws and an uprooting of dominant legal systems and practice, as well as existing power hierarchies. In particular, this potential of legal pluralist forms of riverine rights to transcend and bridge different sets of knowledge and legal orders may be valuable in postcolonial contexts. There is a decolonial promise in this approach that can be realized under favourable circumstances, as the case of Whanganui illustrates.

Yet, the very *idea* of giving rights to a river or other forms of ecosystems is by no means insignificant for the productive potential here. The idea has so far attracted much attention and played a significant role in inspiring new cases of personhood rights for water bodies and the global rights of nature movement. Rights of nature, and of rivers in particular, has a large imaginative force, not to say political force in mobilizing people in and across diverse geographies and lived realities. The underlying idea of bridging the human-nature divide and equating other-than-humans with humans within formal law resonates with how many people experience their relationship to the environment. Moreover, in a time of growing recognition of the urgency of current environmental crises and increasing lived experiences with environmental change, the time is ripe for a radical change in how dominant society thinks about and behaves towards nature. Hence, the broader moral change that rights of nature and rivers invoke has perhaps a more transformative potential and possible wide-reaching effects than the legal change alone.

A related crucial difference between the Whanganui and Atrato on the one hand and the Ganga and Yamuna on the other is that in the former cases rights were given to people and rivers together, whereas in the Indian case, rights were given to the rivers alone. For Whanganui, rights of the river emerged as a consequence of the specific mechanism of realizing Indigenous rights, through legal personhood. In the Atrato case, the concept of biocultural rights explicitly connects the rights of communities and rivers. In India, in contrast, one could argue that rights of nature are explicitly framed as overriding human rights, for instance as the Glaciers ruling includes the injunction to prevent beggars from entering the *ghats* (areas for accessing the river).

Critical commentators have pointed out the risks that certain ways of constructing rights of nature may conflict with human rights and lead to the disenfranchisement of vulnerable groups and Indigenous people (O'Donnell and Macpherson 2018, 7, Marshall 2019, Tanasescu 2022, 118). The India example shows that this risk certainly exists. Yet, the examples from Aotearoa New Zealand and Colombia show that there are ways in which river rights can be framed that have the potential to overcome this possible conflict between rights of nature and rights of people. Throughout this book, we have shown how the cases of granting personhood rights to rivers concern human-river relationships and socio-environmental issues, not just concerns for the river alone. In conclusion, we wish to emphasize this point by arguing that the attention should be placed on *riverine* rights – as opposed to just river rights. By the phrase "riverine" we include those who are related to the river: a wealth of living beings and materials, including (a sometimes diverse group of) local humans who know and care for the river. This implies a perspective that is "more-than-human" but, crucially, also "more-than-river".

We also argue for attention *beyond* riverine rights, so as not to lose sight of the significance of case mandates, governance models and institutional structures, as well as the larger structural preconditions and the social, political and cultural context. More broadly, these cases are not only cases of riverine rights and rights of nature (e.g., about legal rights) but cases of environmental struggles with clear political dimensions. They are all examples of efforts to combat locally specific environmental problems and injustices. In a broader sense, they are practical examples of local efforts to meet the global socio-environmental crises of our time. Even if the direct practical impacts in many cases might be slow, small or non-existent, these efforts are still always contributions towards the (paradigm) shift that is needed for addressing the accumulating environmental crises the world is facing.

Notes

1 *Te Awa Tupua* (Whanganui River Claims Settlement) Act, s. 70d.
Mana is often rendered in English as "power, authority, prestige and effectiveness" (Boyce 2010: 49).
2 Te Awa Tupua (Whanganui River Claims Settlement) Act, s. 16.
3 https://www.ngatangatatiaki.co.nz/our-projects/mouri-turoa/

References

Boelens, Rutgerd, Lena Hommes and Jamie Hoogesteger. 2024. "Ontological politics in river defence debates: Unpacking fields of contention in eco-centric and non-human turns." *Water Alternatives* 17 (3): 649–668.

Boyce, Mary. 2010. "Mana Aha? Exploring the use of Mana in the Legal Māori Corpus." *NZACL Yearbook* 16: 41–61.

Cribb, Miriama, Elizabeth Macpherson and Axel Borchgrevink. 2024. "Beyond legal personhood for the Whanganui River: Collaboration and pluralism in implementing the Te Awa Tupua Act." *The International Journal of Human Rights*: 1–24. doi:10.1080/13642987.2024.2314532.

Cuesta Mosquera, Arlinton, and Catalina Vallejo Piedrahíta. 2024. "Take me to the River: Have riverine rights enhanced community participation in environmental governance within the Atrato River basin?" *The International Journal of Human Rights*: 1–19. doi:10.1080/13642987.2024.2367468.

Finlayson, Christopher, and James Christmas. 2021. *He Kupu Taurangi: Treaty settlements and the future of Aotearoa New Zealand*. Huia Publishers.

Kinkaid, Eden. 2019. "'Rights of nature' in translation: Assemblage geographies, boundary objects, and translocal social movements." *Transactions of the Institute of British Geographers* 44 (3): 555–570.

Macpherson, Elizabeth, and Felipe Clavijo Ospina. 2018. "The pluralism of river rights in Aotearoa New Zealand and Colombia." *Journal of Water Law* 25: 283–293.

Marshall, Virginia. 2019. "Removing the veil from the 'Rights of Nature': The dichotomy between First Nations customary rights and environmental legal personhood." *Australian Feminist Law Journal* 45 (2): 233–248.

McNeish, John A., and Jenny Moreno Socha. 2024. "Muddy waters: On the problematic political ecology of the Atrato ruling, Colombia." *The International Journal of Human Rights*: 1–21. doi:10.1080/13642987.2024.2324121.

O'Donnell, Erin, and Elizabeth Macpherson. 2018. "Voice, power and legitimacy: The role of the legal person in river management in New Zealand, Chile and Australia." *Australasian Journal of Water Resources*: 1–10.

Stone, Christopher. 1972. "Should trees have standing: Toward legal rights for natural objects." *California Law Review*: 450–493.

Tanasescu, Mihnea. 2022. *Understanding the rights of nature: A critical introduction.* Transcript Verlag.

Index